ADVANCES IN EXPERIMENTAL MEDICINE AND BIOLOGY
VOLUME 521

Immune Mechanisms of Pain and Analgesia

Halina Machelska, Ph.D.
Christoph Stein, M.D.
Klinik für Anaesthesiologie und operative Intensivmedizin,
Klinikum Benjamin Franklin, Freie Universität Berlin,
Berlin, Germany

LANDES BIOSCIENCE / EUREKAH.COM
GEORGETOWN, TEXAS
U.S.A

KLUWER ACADEMIC / PLENUM PUBLISHERS
NEW YORK, NEW YORK
U.S.A

Immune Mechanisms of Pain and Analgesia

Advances in Experimental Medicine and Biology Volume 521

Landes Bioscience / Eurekah.com
and
Kluwer Academic / Plenum Publishers
Designed by Celeste Carlton

Printed in the U.S.A.

Kluwer Academic / Plenum Publishers, 233 Spring Street, New York, New York, U.S.A. 10013
http://www.wkap.nl/

Please address all inquiries to Landes Bioscience / Eurekah.com:
Landes Bioscience / Eurekah.com, 810 South Church Street, Georgetown, Texas, U.S.A. 78626
Phone: 512/ 863 7762; FAX: 512/ 863 0081; www.Eurekah.com; www.landesbioscience.com
Landes tracking number: 1-58706-062-0

Immune Mechanisms of Pain and Analgesia, edited by Halina Machelska and Christoph Stein, Landes / Kluwer dual imprint/ Advances in Experimental Medicine and Biology Volume 521, ISBN 0-306-47692-4

Library of Congress Cataloging-in-Publication Data

CIP applied for but not received at time of publication.

ADVANCES IN EXPERIMENTAL MEDICINE AND BIOLOGY

Recent Volumes in this Series

Volume 506
LACRIMAL GLAND, TEAR FILM, AND DRY EYE SYNDROME 3: BASIC SCIENCE AND CLINICAL RELEVANCE
Edited by David A. Sullivan, Michael E. Stern, Kazuo Tsubota, Darlene A. Dartt, Rose M. Sullivan, and B. Britt Bromberg

Volume 507
EICOSANOIDS AND OTHER BIOACTIVE LIPIDS IN CANCER, INFLAMMATION, AND RADIATION INJURY 5
Edited by Kenneth V. Honn, Lawrence J. Marnett, Santosh Nigam, and Charles Serhan

Volume 508
SENSORIMOTOR CONTROL OF MOVEMENT AND POSTURE
Edited by Simon C. Gandevia, Uwe Proske and Douglas G. Stuart

Volume 509
IRON CHELATION THERAPY
Edited by Chiam Hershko

Volume 510
OXYGEN TRANSPORT TO TISSUE XXIII: OXYGEN MEASUREMENTS IN THE 21ST CENTURY: BASIC TECHNIQUES AND CLINICAL RELEVANCE
Edited by David F. Wilson, John Biaglow and Anna Pastuszko

Volume 511
PEDIATRIC GENDER ASSIGNMENT: A CRITICAL REAPPRAISAL
Edited by Stephen A. Zderic, Douglas A. Canning, Michael C. Carr and Howard McC. Snyder III

Volume 512
LYMPHOCYTE ACTIVATION AND IMMUNE REGULATION IX: LYMPHOCYTE TRAFFIC AND HOMEOSTASIS
Edited by Sudhir Gupta, Eugene Butcher and William Paul

Volume 513
MOLECULAR AND CELLULAR BIOLOGY OF NEUROPROTECTION IN THE CNS
Edited by Christian Alzheimer

Volume 514
PHOTORECEPTORS AND CALCIUM
Edited by Wolfgang Baehr and Krzysztof Palczewski

Volume 515
NEUROPILIN: FROM NERVOUS SYSTEM TO VASCULAR AND TUMOR BIOLOGY
Edited by Dominique Bagnard

Volume 516
TRIPLE REPEAT DISORDERS OF THE NERVOUS SYSTEM
Edited by Lubov T. Timchenko

Volume 517
DOPAMINERGIC NEURON TRANSPLANTATION IN THE WEAVER MOUSE MODEL OF PARKINSON S DISEASE
Edited by Lazaros C. Triarhou

Volume 518
ADVANCES IN MALE MEDIATED DEVELOPMENTAL TOXICITY
Edited by Bernard Robaire and Barbara F. Hales

Volume 519
POLYMER DRUGS IN THE CLINICAL STAGE
Edited by Maeda, et al.

Volume 520
CYTOKINES AND CHEMOKINES IN AUTOIMMUNE DISEASE
Edited by Pere Santamaria

Volume 521
IMMUNE MECHANISMS OF PAIN AND ANALGESIA
Edited by Halina Machelska and Christoph Stein

CONTENTS

Preface ix

1. **Glial Proinflammatory Cytokines Mediate Exaggerated Pain States: Implications for Clinical Pain** 1
Linda R. Watkins, Erin D. Milligan and Steven F. Maier
Abstract for General Audience 1
Abstract for Scientific Audience 1
Historical Overview 2
Glia as Modulators of Pain 3
Why does Glial Activation Enhance Pain? 4
Proinflammatory Cytokines in Both Spinal Cord and Brain Influence Pain 11
So, How Do Proinflammatory Cytokines Exaggerate Pain? 12
Clinical Implications and Conclusions 13

2. **Peripheral Hyperalgesic Cytokines** 22
Fernando Q. Cunha and Sérgio H. Ferreira
Hyperalgesia and Classic Inflammatory Mediators 22
Nociceptive Methods and Detection of Hyperalgesic Cytokines 23
Peripheral Hyperalgesia Induced by Cytokines 24
The Indirect Peripheral Hyperalgesic Effects of Cytokines 25
Inflammatory Stimuli and Cytokine Release 27
Bradykinin and Cytokine Release 28
Limitation of the Release and Action of Hyperalgesic Cytokines by Analgesic Cytokines 29
The Cellular Environment and Cytokine Release 31
Cytokines and Peripheral Memory of Hyperalgesia 31
Pharmacological Control of Hyperalgesic Cytokine Action 32
Conclusions 34

3. **Cytokines and Peripheral Analgesia** 40
Michael Schäfer
Introduction 40
Opioid Peptide Release 41
CRF and IL-1 Receptors on Immune Cells 41
CRF- and IL-1-Induced Analgesia 43
Physiological Relevance of CRF- and IL-1-Induced Analgesia 46
Summary 48

4. **Opioid Peptides in Immune Cells** 51
Eric M. Smith
Introduction 51
Leukocyte Production of Opioids 53
In Vivo Leukocyte Opioid Production and Action 58
Other Opioids, Hormones and Cytokines 61
Implications and Future Directions 62

5. **Opioid Receptors on Peripheral Sensory Neurons** ... 69
Christoph Stein
Introduction ... 69
Anatomy ... 69
Electrophysiology ... 70
Alterations During Inflammation ... 71
Analgesic Effects ... 72
Tolerance ... 72
Conclusions ... 73

6. **Morphological Correlates of Immune-Mediated Peripheral Opioid Analgesia** ... 77
Shaaban A. Mousa
Introduction ... 77
Expression of Opioid Receptors on Peripheral Sensory Neurons ... 77
Expression of Opioid Peptides in Immune Cells ... 78
Expression of Corticotropin-Releasing Factor and Interleukin-1 (IL-1) Receptors on Immune Cells ... 81
Expression of Adhesion Molecules ... 82
Clinical Studies ... 82
Summary ... 83

7. **Functional Evidence of Pain Control by the Immune System** ... 88
Halina Machelska
Introduction ... 88
Peripheral Opioid Receptors ... 88
Peripheral Opioid Peptides ... 89
Interactions of Immune-Derived Opioids with Peripheral Opioid Receptors ... 89
Conclusions ... 96

8. **Opioid Receptor Expression and Intracellular Signaling by Cells Involved in Host Defense and Immunity** ... 98
Burt M. Sharp
Abstract ... 98
Introduction ... 98
Identification of Classical and Atypical Opioid Receptors on Immune Cells by Radioligand Binding ... 99
Identification of Opioid Receptor Transcripts in the Immune System ... 100
Regulation of DOR Transcript Expression ... 100
Identification of KOR and DOR by Indirect Fluorescence and Immunofluorescence Labeling ... 101
Opioid Receptor-Mediated Intracellular Signaling in the Immune System ... 102
Summary ... 103

9. **Experimental Evidence for Immunomodulatory Effects of Opioids** **106**
Paola Sacerdote, Elena Limiroli and Leda Gaspani
Introduction 106
Morphine and Endogenous Opioids 106
Modulation of Th1/Th2 Responses 111
Involvement of Opioids in Stress-Induced Immunosuppression 111
Conclusions 112

10. **The Immune-Suppressive Effects of Pain** **117**
Gayle G. Page
Introduction 117
Animal Studies of Pain and Immune Suppression 117
Pain and Immune Function in Humans 118
Animal Studies of Pain, Metastasis and Immune Suppression 118
Discussion and Conclusions 122

11. **Invertebrate Opiate Immune and Neural Signaling** **126**
George B. Stefano, Patrick Cadet, Christos M. Rialas, Kirk Mantione, Federico Casares, Yannick Goumon and Wei Zhu
Introduction 126
The Presence of Opioids and Their Binding Sites 127
Opioid Processing 135
Conclusions 141

12. **Anti-inflammatory Effects of Opioids** **148**
Judith S. Walker
Overview 148
Opioids—Receptor Pharmacology 148
Anti-inflammatory Effects of Opioids 149
Opioids—Peripheral Actions 149
κ-opioids 149
κ-Opioids as Anti-arthritic Agents 150
Mechanisms Responsible for Anti-inflammatory Effects of Opioids 151
Summary and Conclusions 156

Index **161**

EDITORS

Halina Machelska, Ph.D.
Chapter 7

Christoph Stein, M.D.
Chapter 5

Klinik für Anaesthesiologie und operative Intensivmedizin, Klinikum Benjamin Franklin, Freie Universität Berlin, Berlin, Germany

CONTRIBUTORS

Patrick Cadet
Neuroscience Research Institute
State University of New York
Old Westbury, New York, U.S.A.
Chapter 11

Federico Casares
Neuroscience Research Institute
State University of New York
Old Westbury, New York, U.S.A.
Chapter 11

Fernando Q. Cunha
Departamento de Farmacologia
Campus da USP
Ribeirão Preto, SP, Brasil
Chapter 2

Sérgio H. Ferreira
Departamento de Farmacologia
Campus da USP
Ribeirão Preto, SP, Brasil
Chapter 2

Leda Gaspani
Department of Pharmacology
University of Milano
Milano, Italy
Chapter 9

Yannick Goumon
Neuroscience Research Insitute
State University of New York
Old Westbury, New York, U.S.A.
Chapter 11

Elena Limiroli
Department of Pharmacology
University of Milano
Milano, Italy
Chapter 9

Steven F. Maier
Department of Psychology and the Center for Neurosciences
University of Colorado at Boulder
Boulder, Colorado, U.S.A.
Chapter 1

Kirk Mantione
Neuroscience Research Insitute
State University of New York
Old Westbury, New York, U.S.A.
Chapter 11

Erin D. Milligan
Department of Psychology and the Center for Neurosciences
University of Colorado at Boulder
Boulder, Colorado, U.S.A.
Chapter 1

Shaaban A. Mousa
Klinik für Anaesthesiologie und operative Intensivmedizin
Klinikum Benjamin Franklin, Freie Universität Berlin
Berlin, Germany
Chapter 6

Gayle G. Page
Johns Hopkins University School of Nursing
Baltimore, Maryland, U.S.A.
Chapter 10

Christos M. Rialas
Neuroscience Research Insitute
State University of New York
Old Westbury, New York, U.S.A.
Chapter 11

Paola Sacerdote
Department of Pharmacology
University of Milano
Milano, Italy
Chapter 9

Michael Schäfer
Klinik für Anaesthesiologie und operative Intensivmedizin
Klinikum Benjamin Franklin, Freie Universität Berlin
Berlin, Germany
Chapter 3

Burt M. Sharp
Department of Pharmacology
University of Tennessee
Memphis, Tennessee, U.S.A.
Chapter 8

Eric M. Smith
Department of Psychiatry and Behavioral Sciences
University of Texas Medical Branch
Galveston, Texas, U.S.A.
Chapter 4

George B. Stefano
Neuroscience Research Insitute
State University of New York
Old Westbury, New York, U.S.A.
Chapter 11

Judith S. Walker
School of Physiology and Pharmacology
University of New South Wales
Sydney, Australia
Chapter 12

Linda R. Watkins
Department of Psychology and the Center for Neurosciences
University of Colorado at Boulder
Boulder, Colorado, U.S.A.
Chapter 1

Wei Zhu
Neuroscience Research Insitute
State University of New York
Old Westbury, New York, U.S.A.
Chapter 11

PREFACE

Classically, pain sensation or suppression has been attributed exclusively to neuronal circuits. This book challenges and expands this view and offers a critical analysis of a new concept: the contribution of immune mechanisms in pain and analgesia. Among many transmitters with potential for neuro-immune interactions, we concentrate here on those which have been shown to be of functional relevance i.e., cytokines and opioids. Cytokines, low molecular weight proteins produced predominately by inflammatory leukocytes were originally recognized as communication signals between immune cells and as mediators of the host's response to infection. However, they can also influence neural signaling causing either pain or analgesia. Opioid peptides are the natural correlates to morphine and related drugs which remain the major therapy for moderate to severe pain. Extending the traditional view that opioid analgesia arises exclusively within the central nervous system, peripheral opioid receptors can also mediate analgesic effects when activated by exogenous or endogenous opioids within injured tissue. This is of significance because peripherally acting opioids are devoid of centrally-mediated side effects such as respiratory depression, sedation, dysphoria and dependence.

This book presents a comprehensive overview of this emerging and promising area. The first part constitutes an extensive discussion of diverse sensory effects of cytokines. Linda R. Watkins and colleagues provide evidence for exaggerated pain states produced by cytokines derived from glia within the central nervous system. Similar hyperalgesic effects of cytokines in peripheral tissue are analyzed in the chapter by Fernando Q. Cunha and Sérgio H. Ferreira. Their observations are contrasted in the chapter by Michael Schäfer who discusses analgesic actions of cytokines.

The effects of opioids derived from the cells of the immune system are the topic of the next several chapters. The opioid-immune link is highlighted in the chapter by Eric M. Smith who summarizes the poorly understood and controversial issue of the production of opioid peptides by immune cells. Peripheral analgesic effects of immune-derived opioids are the focus of the following chapters. Christoph Stein reviews the anatomy and electrophysiology of opioid receptors localized on peripheral sensory nerves as well as the analgesic effects resulting from the activation of these receptors by exogenous opioid compounds. Shaaban A. Mousa provides a morphological analysis of opioid receptors on sensory nerves and opioid peptides in immunocytes. This chapter is complemented by the chapter by Halina Machelska who describes mechanisms of intrinsic opioid-mediated pain inhibition and emphasizes the clinical relevance of such effects.

The next two chapters examine the influence of opioids on the immune system. Burt M. Sharp deals with the expression and intracellular effects of opioid receptors on immunocytes. The complex issue of immunomodulatory effects of opioids is reviewd in the chapter by Paola Sacerdote and collaborators. The commonly reported immunosuppressive effects of opiates in vitro competes with the observation of immunosuppressive actions of pain itself. This is the topic discussed by Gayle G. Page who suggests that treatment of pain by morphine and other drugs can enhance the protective effects of the immune system.

The chapter by George B. Stefano and colleagues offers a detailed overview on opioid immune-neural communications in invertebrates and documents that these interactions are conserved during evolution. The chapter by Judith S. Walker complements opiate-immune connections presenting opioids as potent anti-inflammatory drugs.

This book is comprised of an internationally recognized group of researchers and presents convincing evidence that cytokines, typical immune products and opioids, traditionally thought to be exclusively of neural origin can integrate immune and nervous system functions. We hope that this book will appeal to basic scientists as well as to clinically-oriented investigators to stimulate future research. We are extremely grateful to all the contributors who made extraordinary efforts to write these overviews of their respective fields.

Halina Machelska
Christoph Stein

CHAPTER 1

Glial Proinflammatory Cytokines Mediate Exaggerated Pain States: Implications for Clinical Pain

Linda R. Watkins, Erin D. Milligan and Steven F. Maier

Abstract for General Audience

When you hurt yourself, you become consciously aware of the pain because a chain of neurons carries the pain message from the injury to the spinal cord, and then from the spinal cord up to consciousness in the brain. However, it has been known for more than two decades that neural circuits within the spinal cord can cause your conscious experience of pain to be amplified—that is, the pain you perceive is out of proportion to the injury that caused it. Until now, all research aimed at understanding how pain amplification occurs in the spinal cord and all drug therapies aimed at curing exaggerated pain have focused exclusively on neurons. This is because neurons were the only type of cell believed to be important in pain. The present review argues that neurons in fact are not the only cell type involved. Rather, that spinal cord cells called "glia" are also critically important. Indeed, when glia become activated, they begin releasing a variety of chemical substances that causes the pain message to become amplified, thus causing pain to hurt more. This review discusses evidence that glia cause pain to become amplified and describes how the glia cause this to happen. The take-home message is that drugs that target glia and the chemical substances that these glia release are predicted to be powerful remedies for pain problems in people.

Abstract for Scientific Audience

It is now clear that pain can be dynamically modulated by spinal circuitry so to create either pain suppression or pain enhancement. Pain enhancement can be in the form of lowered threshold for responding to either thermal stimuli (thermal hyperalgesia) or touch/pressure stimuli (mechanical allodynia). Classically, pain modulatory circuits have been conceptualized as being composed solely of neurons. Here we refute that view. Instead, we review evidence that enhanced pain states may be created and/or perseverated in spinal cord by products released by activated glia. Some glial products are identical to those classically associated with pain enhancement. Others, such as proinflammatory cytokines (interleukin-1, tumor necrosis factor, and interleukin-6), are newly recognized mediators of both thermal hyperalgesia and mechanical allodynia. Indeed, there is growing evidence

Immune Mechanisms of Pain and Analgesia, edited by Halina Machelska and Christoph Stein.

that spinal proinflammatory cytokines are key mediators of exaggerated pain states that are created following either subcutaneous, intraperitoneal, peri-sciatic, or intrathecal administration of various immune activators. It is clear from the studies reviewed that spinal cord glia and glial proinflammatory cytokines appear to be excellent targets for pain control.

This chapter will be different in focus than most others in this volume. Here, the focus will be not on analgesia, but rather on exaggerated pain states. Further, the principal focus will not be on immune cells in the periphery, but rather on immune-like cells within the central nervous system: microglia and astrocytes. These cells will be collectively referred to as glia. The argument that will be developed is simply stated as follows: spinal cord glial activation provides a major driving force for creating and maintaining a wide variety of exaggerated pain states. This is a new view of pathological pain, and one that suggests that developing drugs to target glial function may provide new approaches to clinical pain control.

Historical Overview

Conceptualization of pain has undergone metamorphosis over the centuries, from Aristotle's view of pain being a "passion of the soul" to the mid-1900s view of pain as a "labeled line" pathway from periphery to consciousness with faithful reproduction of sensory stimuli at every level of neural processing. This "labeled line" pathway was conceived of as a chain of synaptically connected neurons from the periphery to cortex. Glia had no role in this schema, as they lacked axons and so were not thought of in terms of cell-to-cell signaling.

Modern views of pain recognize that pain is far more complex than a simple "labeled line". Pain is now recognized to involve the interplay of sensory, evaluative, and affective components to create the pain experience. But even here, pain normally begins with transmission of the stimulus event from the periphery to the spinal cord. This first step in central pain transmission is now known to be dynamically regulated with the pain signal capable of being either suppressed or exaggerated by circuitries within the spinal cord. Further modulations of the pain message can occur as the information is sent from the spinal cord up to the brain, to consciousness. However, from sensation to perception, pain pathways and the circuits that modulate the pain message are still classically viewed as entirely composed of neurons.

Pain suppression circuitries were the first of the pain modulatory systems to be discovered.[1] Activated by opiates, electrical stimulation of discrete brain regions, and environmental stressors, these neural circuits have been well mapped as to their pathways and neurochemistries.[1,2] These have been proposed to have evolved to serve a survival function in times of fight/flight, allowing the organism to be oblivious to pain so as to maximize the chance of successful defense or escape.[3] Multiple pain suppressive circuits have now been identified. All of these circuits have been proposed to be chains of neurons. These neuron chains arise, as one example, in the periaqueductal gray, which in turn excite neurons in the ventral medial medulla, which then inhibit pain transmission neurons in the spinal cord dorsal horn.[1,2] While there is some scattered evidence that glia may be involved in pain suppression[4], they will not be the focus of the present discussion.

Instead, the focus will be on pathways that enhance pain. Pain enhancement circuitries have been a major focus of pain research in recent years. These act to amplify the pain signal that is sent by spinal cord neurons toward consciousness. Behaviorally, this is observed as a lowering of pain threshold such that the organism now responds more rapidly to a thermal stimulus (thermal hyperalgesia) and/or withdraws from light touch/pressure stimuli that go unnoticed by normal animals (mechanical allodynia). Pain enhancement circuits that create thermal hyperalgesia and mechanical allodynia can be activated by inflammation, infection or damage that occurs in either the periphery or the central nervous system.[5] These circuits can also be accessed by activating intraspinal circuits or discrete brain-to-spinal cord pathways. While the spinal circuits have long been the focus of study, the brain-to-spinal cord circuits are still being explored as to their pathways and neurochemistries.[6] Although pain enhancement circuits were originally conceived of as purely composed of neurons, this view is now changing. The reasons for this fundamental change will be reviewed below.

Glia as Modulators of Pain

In the early 1990s glia came to the attention of pain researchers. By this time, it had been documented that damage of peripheral nerves causes intense microglial and astrocyte activation in the central nervous system.[7-9] Glial activation occurs specifically in the neural region containing central terminals and/or somas of the damaged peripheral nerves. Activation of these glia is readily observable since activated astrocytes increase their expression of glial fibrillary acidic protein (GFAP) and microglia increase their expression of complement type 3 receptor, and both GFAP and complement type 3 receptor expression are easily detected by immunohistochemistry.[8,10,11]

This observation that glia are activated by peripheral nerve trauma is intriguing since such injuries also cause exaggerated pain states, called neuropathic pain. Garrison et al [11] used the sciatic chronic constrictive injury (CCI) neuropathic pain model to test whether elevated GFAP levels in spinal cord dorsal horn would correlate with neuropathic pain measured in the same animals. A strong correlation was found. Garrison et al[12] further found that peri-spinal (intrathecal; i.t.) administration of an N-methyl-D-aspartate (NMDA) receptor antagonist (MK-801) blocked both the neuropathic pain behavior and elevations in GFAP. Thus astrocyte activation appeared, at minimum, to be strongly correlated with the expression of neuropathic pain behaviors.

These findings led Meller et al[13] to test whether spinal cord glial activation is necessary and sufficient to produce enhanced pain. Disruption of spinal cord glial function with a glial metabolic inhibitor was found to reduce both thermal hyperalgesia and mechanical allodynia induced by peripheral inflammation using s.c. zymosan (yeast cell walls). To test whether glial activation was sufficient to enhance pain responses, Meller et al[13] took advantage of the fact that glia act like immune cells. That is, glia become activated upon binding to foreign substances such as bacterial cell walls. Indeed, immune activation of glia by peri-spinal injection of lipopolysaccharide (a constituent of the cell walls of gram negative bacteria) created exaggerated pain responses.[13]

Microglia were first implicated in exaggerated pain by Watkins et al.[14] Following up on the work of Garrison et al[11,12] and Meller et al[13], Watkins et al[14] reported that thermal hyperalgesia induced by either intraperitoneal (i.p.) bacteria or subcutaneous (s.c.) formalin injection was correlated with immunohistochemical evidence of microglial as well as astrocyte activation within the spinal cord dorsal horns.

Spinal cord dorsal horn astrocytes and/or microglia have now been reported to be activated in response to a wide array of conditions known to produce exaggerated pain responses. These include peripheral inflammation from s.c. formalin[14-16], s.c. zymosan[15], i.p. bacteria[14]; peripheral nerve trauma[15,17-20], bone cancer[21]; lumbar root constriction[22]; spinal nerve transection[19]; spinal cord trauma[23]; activation of brain-to-spinal cord pain facilitatory circuits[14,24,25]; and spinal cord immune activation.[26] Glial activation appears to create and/or maintain exaggerated pain states since pharmacological disruption of spinal cord glial function disrupts pain induced by s.c. zymosan[13], s.c. formalin[27], peripheral nerve inflammation[28-30], spinal nerve transection[31], and spinal cord immune challenges.[26,32] To date, comparable immunohistochemical or glial functional disruption studies have not been performed in brain.

Why does Glial Activation Enhance Pain?

Glial Regulation of Substances that Classically Mediate Exaggerated Pain States

Astrocytes and microglia release a variety of substances upon activation. The exact substances released by each of these cells are not identical, as each cell type has unique characteristics and functional capabilities.[33-35] However, for the purpose of the present discussion, microglia and astrocytes will be treated as a unit since they typically are co-activated by pain-inducing stimuli.

Microglia and astrocytes can vary the type, amount, and temporal pattern of released substances dependent upon a variety of factors. These include which neurotransmitters have been released in their vicinity, which bacteria/viruses are present, whether inflammatory mediators are present, prior history of activation, etc. Given this complexity and the recency of interest in spinal cord glia, it is not surprising that investigation of glial modulation of pain is still in its infancy.

Microglia and astrocytes are attractive candidates as mediators of exaggerated pain states for a number of reasons. First, upon activation, they release a wide variety of substances known to excite pain transmission neurons including (a) nitric oxide (NO), superoxide, and other free radical species; (b) prostaglandins, leukotrienes, and arachidonic acid; (c) chemokines and proinflammatory cytokines, including interleukin-1β (IL-1), tumor necrosis factor-a (TNF), and IL-6; (d) glutamate and other excitatory amino acids; and (e) nerve growth factors.[33-35] Of these, proinflammatory cytokines will be the focus of a separate section, below. Second, of the factors released by activated glia, IL-1, prostaglandins, and NO each cause exaggerated release of substance P from primary afferents.[36-39] Third, microglia and astrocytes are ideal candidates for driving perseverative changes characteristic of exaggerated pain states. This is because microglia and astrocytes form positive feedback loops in which substances released by one cell type further activate the other, causing prolonged release of excitatory substances from these cells.[33,35] Fourth, the substances released from glia act in an autocrine/paracrine fashion, and so are ideal for activating both glia and neurons in their general vicinity.[33,35] Indeed, glially-released substances diffusing to excite a larger spinal population of neurons and glia could easily contribute to the classically observed temporal expansion in the body region experiencing enhanced pain. Fifth, microglia and astrocytes outnumber neurons about 10:1, so activation of these cells could be expected to significantly impact the function of neurons in the area. Sixth, these glial cells are activated by neurotransmitters (substance P, glutamate, aspartate, etc.) released by small diameter sensory afferents activated by bodily injury, nerve damage, infection or inflammation.[40,41] And seventh, these glial cells are also activated by immune challenges (trauma, viruses, bacteria, etc.) of the central nervous system that are associated with exaggerated pain states.[26,42-45]

What should be immediately apparent from this discussion is that glia can increase the exposure of pain transmission neurons to substance P, glutamate, aspartate, NO (via glial constitutive as well as inducible NO synthases), prostaglandins, and so forth. These pain-enhancing substances have been the focus of intense study by pain researchers for well over two decades. The fact that glia are now known to dramatically regulate these key factors suggests that at least some of the existing literature on substance P, excitatory amino acid, NO, and prostaglandin involvement in exaggerated pain states may actually reflect an unrecognized contribution of glia to the effects observed.

Glial Release of Proinflammatory Cytokines: Newly Recognized Mediators of Exaggerated Pain States

Beyond classic pain enhancing substances, glia also release proinflammatory cytokines. The proinflammatory cytokines (TNF, IL-1 and IL-6) were given this label because they are involved in orchestrating the early immune response to infection, inflammation, and injury. These cytokines are synthesized and released in the central nervous system as well as in the periphery, and both neurons and glia express receptors for them.[46-49] Regarding their pain modulatory actions, TNF[4,17,26,50-52], IL-1[15,26,27,53-56] and IL-6[4,57] have all been implicated in creating exaggerated pain states by actions within the brain or spinal cord.

Two examples of proinflammatory cytokine-mediated exaggerated pain states will be discussed in detail below. The first will be pain facilitation created by immune activation in spinal cord. The second will focus on pain facilitation created by immune activation around otherwise healthy peripheral nerve trunks. A summary of spinal cord and brain cytokine involvement in pain and clinical implications will then follow.

Immune Activation in Spinal Cord Creates Exaggerated Pain via Release of Spinal Proinflammatory Cytokines

We initially became interested in spinal cord glia as a natural outgrowth of our work on bi-directional immune-brain communication[58] and sickness-induced hyperalgesia[30,59,60], as the immune challenges we commonly used in those studies (i.p. lipopolysaccharide, i.p. IL-1) were known to activate glia within the central nervous system.[61] This led us to discover that spinal cord microglia and astrocytes are indeed activated following either i.p. bacterial challenge or s.c. formalin inflammation.[14] Furthermore, dorsal spinal cord IL-1 protein levels rapidly increase after i.p. immune activation (Fig. 1). Supporting the idea that this spinal glial activation mediates thermal hyperalgesia that follows s.c. formalin, we found that this thermal hyperalgesia was abolished by i.t. administered drugs that either disrupt glial function or block IL-1 receptors.[27]

While glia are clearly implicated as mediators by these experiments, neither the s.c. formalin nor i.p. lipopolysaccharide paradigm allows for selective activation of glia. This is because the glia are only activated secondarily to sensory afferent activity. Thus, central nervous system neurons as well as glia become activated in response to the hyperalgesia-inducing event. It is necessary to test the effect of direct and selective activation of these cells in order to define whether glia themselves are capable of driving exaggerated pain states. We chose to take advantage of the fact that microglia and astrocytes are similar to peripheral immune cells in that they recognize, and respond to, viruses. Thus, we tested the effect of peri-spinal (i.t.) administration of gp120, a glycoprotein expressed on the external surface of the human immunodeficiency virus type 1 (HIV-1). gp120 provides an excellent means of selective activation of astrocytes and microglia since these cells, but not neurons, express so-called activation receptors selective for gp120 (for discussion, see ref. 44). These activation receptors are distinct from the receptors used by

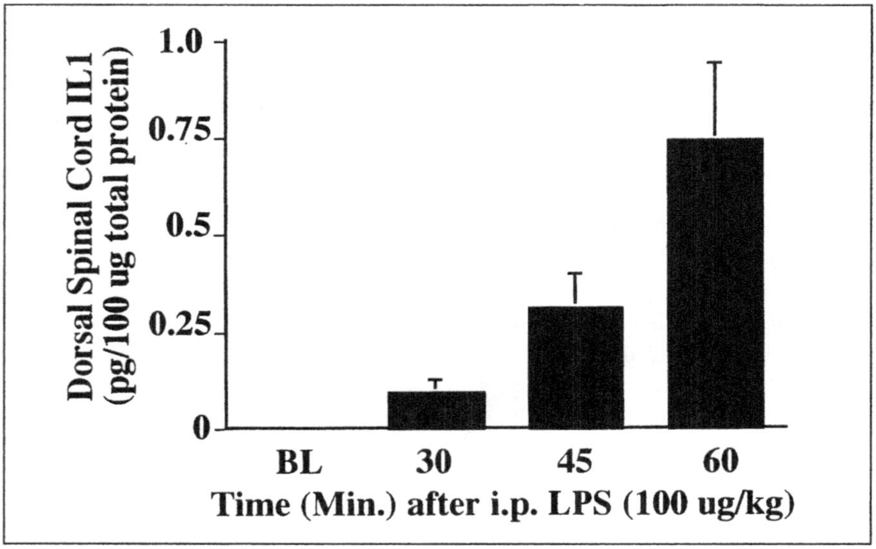

Fig. 1. Enzyme-linked immunosorbant assay (ELISA) data demonstrating that interleukin (IL)-1 protein level in dorsal spinal cord rapidly increases after i.p. administration of a dose of lipopolysaccharide (LPS) that induces thermal hyperalgesia.

HIV-1 to infect cells. Rather, these receptors trigger microglia and astrocytes to begin to produce and release proinflammatory cytokines[62,63], NO[62], excitatory amino acids[64], prostaglandins[65], and so forth.

Rats rapidly (less than 20 min) develop both thermal hyperalgesia and mechanical allodynia upon administration of gp120 over lumbar cord.[26,42] Both pain states resolve after about 3-6 hours. This likely reflects the instability of gp120 produced by its delicate 3-dimensional conformation that is both critical to its pain-inducing effects and easily disrupted by denaturation.[42] gp120-induced thermal hyperalgesia and mechanical allodynia appear to involve spinal cord glia since these pain effects are abolished by either (a) i.t. fluorocitrate, a metabolic inhibitor that preferentially affects glia[42] or (b) i.t. or i.p. CNI-1493, an inhibitor of glially expressed p38 mitogen activated protein (MAP) kinase.[42] Furthermore, immunohistochemical analyses of specific activation markers indicate that i.t. gp120 activates both microglia and astrocytes.[26]

Since these early indications pointed to glia as probable mediators, we then sought to identify which glial products were of importance in mediating exaggerated pain. Since we previously found that spinal cord IL-1 was critical for the glially mediated pain facilitation produced by s.c. formalin[27], we tested whether IL-1 receptor antagonist (IL-1ra) would affect the pain facilitatory effects of i.t. gp120. Indeed, i.t. IL-1ra abolished both i.t. gp120-induced thermal hyperalgesia and mechanical allodynia.[26,30] Furthermore, using a rat-specific enzyme-linked immunoassay (ELISA) for IL-1 protein, we documented that gp120 injected i.t. over lumbar spinal cord rapidly induced the production and release of IL-1.[26,30] Documentation of the release of proinflammatory cytokines is critical for concluding that cytokine elevations are physiologically meaningful, as proinflammatory cytokines can accumulate intracellularly without being released.[66] Hence, the observation that extracellular IL-1 levels increase after gp120 documents that IL-1 could affect the function of other cells in the region. Elevations in dorsal spinal cord and cerebrospinal fluid (CSF) IL-1 levels were site specific, in that no elevations in cervical spinal cord, cervical CSF or peripheral blood IL-1 were found.[26,30]

A logical question that arises at this point is the source of the spinal cord IL-1 that creates exaggerated pain. To begin to address this, we used single-label immunohistochemistry to compare dorsal horn expression of IL-1, GFAP (which selectively labels astrocytes) and OX42 (which selectively labels microglia) in separate tissue sections (Gaykema, Milligan, Maier and Watkins, unpublished observation). What we found was that cells that express IL-1 are strikingly similar in their morphology to astrocytes (Fig. 2). This conclusion was the same regardless of whether the rats were injected with gp120 or not. That is, after gp120, there was no change in the cell type expressing IL-1. As a more rigorous test, we used double-label immunofluorescence to examine for co-expression of IL-1 with either GFAP or OX42 (Gaykema, Milligan, Maier and Watkins, unpublished observation). IL-1 was labeled with a red fluorescent marker. GFAP and OX42 (in separate sections) were labeled with a green fluorescent marker. Co-localization was observed only in astrocytes (Fig. 3). This was again true both basally and following i.t. gp120. Hence, evidence to date points to spinal cord astrocytes as the source of IL-1 mediating exaggerated pain states.

IL-1 is not the only proinflammatory cytokine implicated in the pain enhancing effects of i.t. gp120. Indeed, TNF is just as rapidly detected in lumbosacral CSF and a TNF functional antagonist (TNF binding protein; TNFbp; also known as soluble TNF receptor) attenuates the effect of i.t. gp120 as well.[26,30] The actions of TNF may well be largely indirect, as TNFbp also blocks both production and release of IL-1.[26,30]

How IL-1 and TNF create exaggerated pain states is currently being investigated. As noted above, i.t. and i.p. CNI-14932 completely block both mechanical allodynia and thermal hyperalgesia. This effect could be due to disruption by CNI-1493 of either (a) IL-1 and TNF synthesis or (b) IL-1 and TNF downstream actions. While some immune activators do stimulate IL-1 and TNF production and release via p38 MAP kinase pathways[67-69], gp120 is not such a stimulus.[70] In support of this, no decrease in gp120-induced IL-1 or TNF release into lumbosacral CSF occurs after CNI-1493 pretreatment.[71] Therefore, CNI-1493 must be exerting its effects downstream of this step. In support of this idea, IL-1 and TNF induced effects on NO production, prostaglandins, and excitatory amino acid regulation have each been linked to p38 MAP kinase cascades.[68, 72, 73] Indeed, we have preliminary evidence that NO is a key mediator in gp120 enhanced pain, as a broad-spectrum NOS inhibitor (L-NAME) blocks gp120 induced allodynia.[30,74] Tests of additional putative mediators are underway.

Immune Activation Near Peripheral Nerves Creates Exaggerated Pain via Release of Spinal Proinflammatory Cytokines

We have also been studying a second, and very different, model for examining the pain modulatory effects of central proinflammatory cytokines. Here, the immune activator is not applied to the spinal cord. Rather, immune activation is created around a small portion of a single healthy peripheral nerve trunk. We arrived at this model based on a variety of evidence that neuropathic pain may not simply result from physical trauma to nerves. Rather, as we have reviewed previously, a component of neuropathic pain may arise from immune activation.[60] This is because local immune activation will invariably follow physical trauma to any body tissue, including nerves. Intriguing studies by Maves et al[75] and Clatworthy et al[76-78] pointed to immune activation near peripheral nerves as inducing behavioral and electrophysiological indices of enhanced pain. In addition: (a)

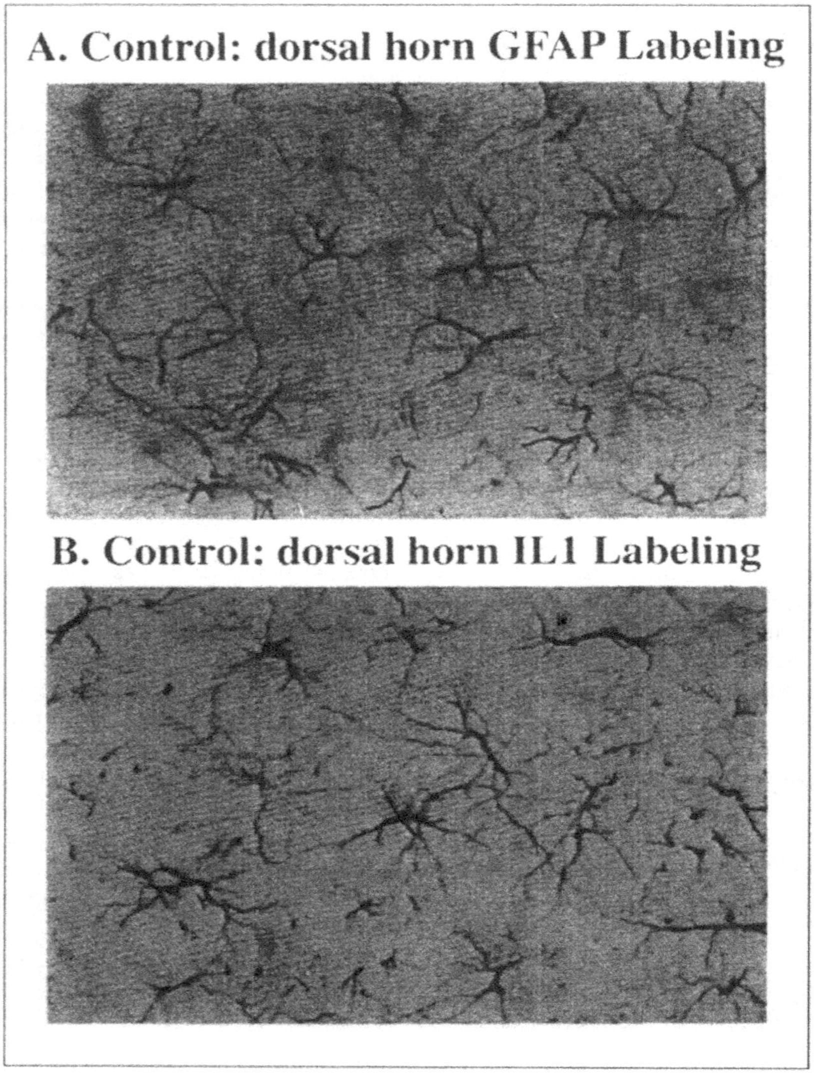

Fig. 2. Single label immunohistochemistry showing strikingly similar morphology of glial fibrillary acidic protein (GFAP)-expressing astrocytes (panel A) and of interleukin (IL)-1-expressing cells (panel B). This strikingly morphological similarity also occurred after i.t. gp120. That is, no new cell type began expressing IL-1 after i.t. gp120, suggesting astrocytes as the source of IL-1 mediating gp120-induced exaggerated pain states.

increases in immune-derived proinflammatory cytokines occur in damaged peripheral nerves[79-81], (b) the timecourse of these cytokine elevations parallel the observation of neuropathic pain[79,82], (c) direct administration of proinflammatory cytokines[83,84] or bacterial components[76-78,85] to peripheral nerves increase behavioral and electrophysiological indices of pain, and (d) blocking either immune activation or the action of either IL-1 or TNF blocks behavioral and electrophysiological evidence of enhanced pain.[82,86-89]

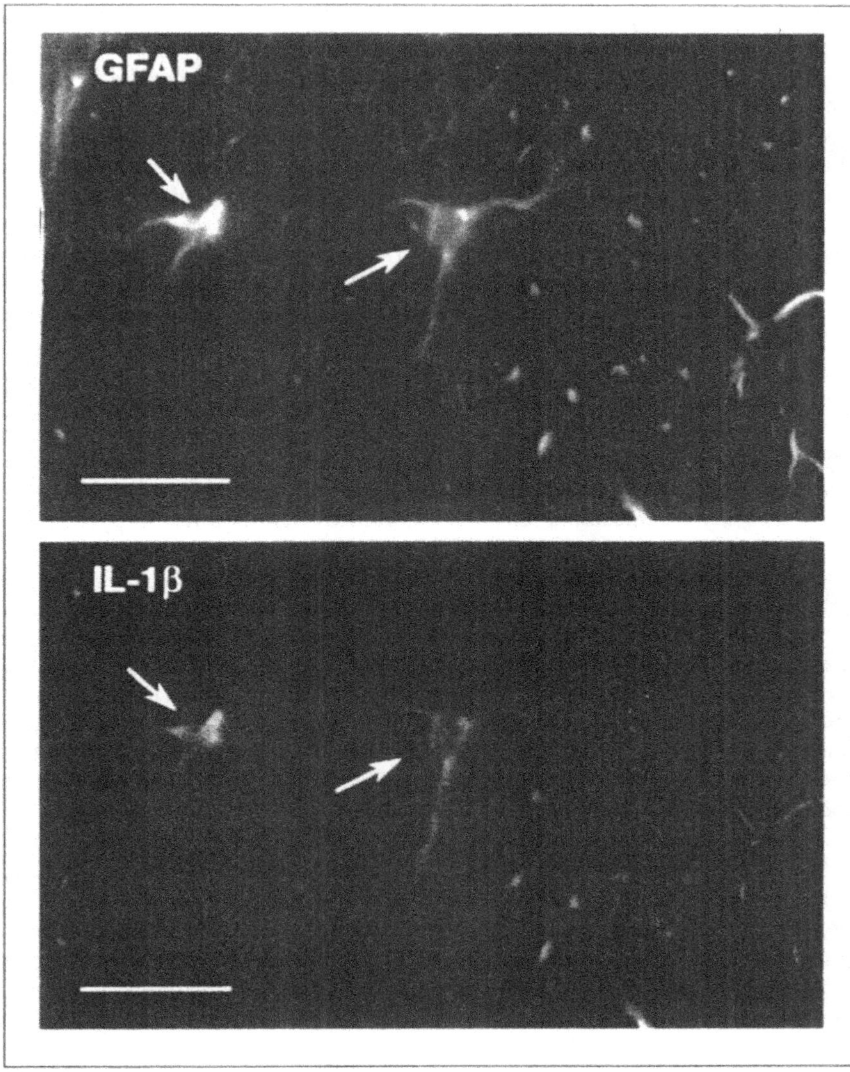

Fig. 3. Double label immunofluorescence showing co-localization of glial fibrillary acidic protein (GFAP); labeled by a green fluorescent marker selective for astrocytes; top panel) and interleukin (IL)-1 (labeled by a red fluorescent marker selective for IL-1; bottom panel). Arrowheads note cell profiles that double-labeled for GFAP and IL-1. Scale bars in both panels = 25 mm.

While other groups have focused on changes in the peripheral nerve created by such immune activation, our laboratory has instead focused on the spinal cord changes created by peri-sciatic immune activation. To create immune activation around a healthy sciatic nerve, we developed procedure in which a soft piece of gelatin (Gelfoam) is wrapped around a small portion of the left sciatic nerve at mid-thigh level. The wrap is attached to a silastic catheter that exits the animal. An advantage of this procedure is that it allows the animal to fully recover prior to testing. This is important since anesthesia can seriously disrupt normal immune function.[90] Indeed, we have found that the pain enhancing effects of zymosan are greatly delayed and attenuated if zymosan is applied to the nerve during surgery rather than after a several day recovery period.[29]

Using this model, we have tested the pain enhancing effect of yeast cell walls (zymosan) injected around the left sciatic nerve. Since nerves don't express receptors for yeast, this allows selective immune activation to be studied. We refer to this localized immune activation as Sciatic Inflammatory Neuropathy (SIN). What we observed after left peri-sciatic zymosan is a rapid (less than an hour) and prolonged (over a week) mechanical allodynia. No thermal hyperalgesia is observed even after a wide range of zymosan doses.[28-30] This effect is not restricted to zymosan as peri-sciatic injection of a proinflammatory cytokine (HMG) also dose-dependently produces unilateral and bilateral allodynia.[28]

Intriguingly, the pattern of mechanical allodynia created by SIN radically changes in a dose-dependent fashion. After low doses (4 μg) of zymosan, a unilateral allodynia is observed, affecting only the zymosan-injected left leg. No pain change occurs in the right, uninjected leg. Bilateral allodynia results at high doses (40-160 μg). While the simplest explanation for SIN-induced bilateral allodynia would have been that zymosan simply reached the systemic circulation, four lines of evidence argue against this possibility.[28-30] First, dye injected into the gelfoam remains in the gelfoam. Second, zymosan injected into gelfoam implanted in immediately adjoining muscle fails to produce allodynia despite the fact that the zymosan has an equal likelihood of reaching the systemic circulation. Third, if zymosan-induced bilateral allodynia were caused by systemic distribution of the immune activator, then a whole body allodynia would be expected. This does not occur. Indeed, no changes in responses to touch/pressure stimuli are observed in front paws even after a wide range of zymosan doses (4-400 μg) injected around the sciatic nerve. Lastly, if zymosan were escaping the gelfoam, enlargement of local, and possibly distant, lymph nodes would be expected. No lymph node enlargement was found. Thus, taken together, the best evidence to date is that SIN-induced ipsilateral and bilateral allodynia must somehow both result from immune activation in the vicinity of a single sciatic nerve.

The fact that strong, unilateral peri-sciatic immune activation leads to allodynia in both the injected (left) and noninjected (right) legs is strikingly similar to human reports of "mirror pain". In mirror pain, a physical cause for pain can easily be found in one body region. However, its mirror image location on the contralateral body surface also gives rise to pain, despite the fact that no pathology exists on that side[91] Long relegated to psychiatrists to treat such "physically impossible" pains, it is now recognized that mirror pain can result from activation of endogenous pain facilitatory circuits. Both intraspinal and brain-to-spinal cord circuits have been proposed.[6,92-94] However, a glial basis for mirror pain has not been considered. Further, nothing is yet known regarding the neuroactive substances that create this phenomenon. Thus the reliable observation of robust mirror pain in this SIN model provides an exciting opportunity to explore the mechanisms underlying such effects.

An additional intriguing aspect of SIN-induced allodynia is that the allodynia extends beyond the skin innervation zone of the sciatic nerve; that is, beyond the body "territory" innervated by sciatic fibers.[28-30] Allodynia is observed in skin innervated by the saphenous nerve as well. This allodynia arising from the experimentally naive saphenous innervation region is called "extra-territorial", referring to the fact that allodynia is induced in skin regions beyond that innervated by the experimentally manipulated sciatic nerve. Given that: (a) the saphenous nerve (which runs along the inner thigh) is physically quite distant from the sciatic nerve (which runs along the outer thigh), and (b) there are no synapses connecting these nerves in the periphery, this implies that zymosan-induced sciatic activity must somehow alter the processing of saphenous information within the spinal cord itself. When unilateral (left leg) allodynia occurs in the sciatic, unilateral (left leg) allodynia occurs in the saphenous cutaneous territory as well. Bilateral saphenous effects parallel bilateral sciatic effects. The fact that the spinal cord dorsal horn termination zone for saphenous nerve fibers is anatomically enveloped by the spinal cord termination zone for sciatic nerve fibers[95] may provide the underlying explanation for this extra-territorial allodynia. That is, substances released in the dorsal horn by immune-excited sciatic fibers may act in a paracrine fashion to evoke exaggerated spinal cord responses to neighboring saphenous inputs as well. Clearly, given the discussion to this point, substances released by activated glial could serve this function well.

Indeed, spinal cord glial activation again appears to underlie SIN-induced exaggerated pain states. Both unilateral and bilateral mechanical allodynias are prevented and/or reversed by i.t. administration of either: (a) a drug that disrupts glial function (fluorocitrate), (b) CNI-1493 (the p38 MAP kinase inhibitor described previously), or (c) IL-1 receptor antagonist.[28-30]

Proinflammatory Cytokines in Both Spinal Cord and Brain Influence Pain

It is clear that proinflammatory cytokines are produced by microglia and astrocytes. Cytokines can be produced by neurons under some conditions as well.[17,96,97] In spinal cord, one very intriguing twist is that the immunohistochemically documented increases in IL-1 protein in neurons occur predominantly, if not exclusively, within the nucleus.[15,17] Nuclear localization of IL-1 is puzzling since cytokines are proteins synthesized in the cytoplasm. One potential explanation of this finding lies in the fact that, upon binding of IL-1 to its functional receptor (IL-1 receptor type I; IL-1RI), the IL-1-IL-1RI complex internalizes.[98-100] Translocation of the chemically intact IL-1 (and hence, immunohistochemically reactive) with its bound receptor to the nucleus has been observed by several laboratories.[100] This complex has been posited to serve as a transcription factor for regulating gene activation.[100] Although IL-1 is not commonly thought of as transducing signals in this fashion in peripheral immune tissues[101], these findings may point to the need for studies that explicitly examine spinal cord neurons.

Spinal cord proinflammatory cytokines show rapid and prolonged elevation under a variety of conditions known to be associated with exaggerated pain states. Spinal cord IL-1 mRNA and/or protein levels rise in response to s.c. inflammation (formalin or zymosan)[15], peripheral nerve injury[17], spinal nerve constriction in a lumbar radiculopathy model[22], spinal cord contusion[102,103,105], and spinal immune activation.[26,30] Blocking spinal IL-1 receptors inhibits nociceptive behavioral responses to s.c. formalin in mice[106], s.c. formalin induced thermal hyperalgesia in rats[27], peripheral nerve inflammation[28-30], and i.t. dynorphin-induced allodynia.[107] Lastly, spinal administration of IL-1 has been reported to elicit pain behaviors, at least in mice.[106] While we did not observe changes in tailflick to radiant heat following i.t. IL-1[56], Meller et al[13] observed enhancement of both thermal and mechanical responsivity following combined i.t. administration of IL-1 plus interferon-gamma.

TNF mRNA[103,108,109] and TNF protein[17,18,26,30,104] are also constitutively expressed in spinal cord. TNF mRNA and/or protein levels have been reported to rise in response to sciatic CCI coincident with pain behaviors[17,104], other forms of peripheral nerve trauma[18], spinal cord contusion[103], and spinal immune activation.[26,30] Sciatic CCI induced allodynia is also exaggerated in mice whose astrocytes have been genetically engineered to overexpress TNF.[51]

IL-6 mRNA differs from IL-1 and TNF in that it does not appear to be constitutively expressed in spinal cord.[108] Rather, in keeping with classical cytokine cascades, IL-6 production and release is delayed relative to IL-1 and TNF. IL-6 mRNA and/or protein levels become elevated in response to peripheral nerve injury[57,110] and spinal cord contusion injury[111]; no other animal models have yet been tested. In humans, lumbosacral CSF levels of IL-6 increase in humans in response to hip replacement surgery. This rise occurs independent of plasma IL-6 levels, supporting a spinal origin for the CSF effects.[112] I.t. IL-6 induces both mechanical allodynia and thermal hyperalgesia[57], whereas IL-6 knockout mice show suppressed pain behaviors to sciatic CCI.[113,114]

Far less is known about pain modulatory effects of proinflammatory cytokines in brain. What is known is that low doses of intracerebroventricular (ICV) IL-1[53-55], TNF[115] and IL-6[116] can cause exaggerated pain states on both thermal and paw pressure tests. Electrophysiology of trigeminal dorsal horn pain transmission neurons reveals that ICV IL-1 specifically enhances neuronal responses to pain stimuli.[54] The effects of the ICV-delivered proinflammatory cytokines appear to be dose-dependent in that higher doses of either IL-1[117] or TNF[52,115,118] can cause analgesic responses, rather than hyperalgesia. Another variant IL-1α, has been reported to produce either no effect[119] or analgesia[118,120-122] Further, TNF protein has been reported to increase in various brain regions following sciatic CCI[104], and CCI-induced thermal hyperalgesia is blocked by ICV anti-TNF antiserum.[52]

Microinjections of proinflammatory cytokines into discrete brain nuclei have revealed site specific effects. Dependent upon the site injected, either pain enhancing or pain suppressive effects are induced. Hyperalgesia has been reported following microinjection of IL-1 into hypothalamus, preoptic area, paraventricular nucleus, and the diagonal band of Broca.[55,123] Analgesia, on the other hand, has been reported (using the same dose ranges as for hyperalgesia in other brain sites) from injections into the hypothalamus, ventromedial hypothalamus, thalamus, centro-medial nucleus, and gelatinous nucleus.[55,123] Such findings clearly argue for the study of cytokines microinjected into specific targets so to avoid simultaneously affecting analgesia- and hyperalgesia-inducing brain regions.

So, How Do Proinflammatory Cytokines Exaggerate Pain?

There are a variety of ways in which cytokines could create exaggerated pain. One way would be by direct actions on neurons. Neurons express receptors for proinflammatory cytokines.[124] Indeed, as noted above, IL-1 rapidly causes pain-specific increases in the excitability of trigeminal dorsal horn neurons.[54] No other proinflammatory cytokine has yet been electrophysiologically tested in either trigeminal or spinal cord dorsal horns.

A second way that proinflammatory cytokines could enhance pain is by modulating release of neurotransmitters from primary afferent fibers. Most studies have focused on substance P. Low doses of IL-1 enhance, while high doses of IL-1 inhibit, substance P release from rat spinal cord slices.[36] IL-1 has also been shown to evoke release of substance P from cultured rat dorsal root ganglion cells, an effect blocked by IL-1 receptor antagonist.[37] In support of the idea that IL-1-induced substance P release is involved in hyperalgesia phenomena, Tadano et al[106] have reported that i.t. IL-1α elicits substance P-like scratching, biting, and licking behavior that is blocked by either IL-1 receptor antagonist or antiserum against substance P. Furthermore, s.c. formalin pain changes are blocked both by i.t. IL-1 receptor antagonist and by i.t. selective antagonists of substance P binding to the high affinity NK-1 receptor.[27,125]

Modulation of substance P release may be important for altering glial, as well as neuronal, activity. Clearly, enhanced substance P release would directly increase the activity of spinal cord pain transmission neurons. However, astrocytes would also be affected. Spinal cord astrocytes express high densities of the high affinity NK-1 receptor that binds substance P.[126] Indeed, spinal astrocytes are far richer in their expression of this receptor than their brain counterparts.[126] Activation of these spinal cord astrocyte NK-1 receptors induces, or potentiates, release of IL-6[126-128], TNF[129] and IL-1[130] from these cells.

Finally, proinflammatory cytokines could modulate pain indirectly. That is, proinflammatory cytokines could alter pain transmission via cytokine-induced release of some other neuroactive substance. Proinflammatory cytokines can induce the release of NO and other oxygen radicals, prostaglandins, excitatory amino acids, and so forth from microglia and astrocytes[33-35], substances that could be the proximate cause for pain enhancing effects of proinflammatory cytokines.

Clinical Implications and Conclusions

The implications of the studies summarized above are clear. This body of evidence suggests that the activation of glia in the spinal cord and the proinflammatory cytokine release that results, drives exaggerated pain states. Indeed, it is quite striking that spinal glial activation and proinflammatory cytokine release are key mediators of exaggerated pain states ranging from acute inflammation to chronic nerve trauma. Given such pervasive involvement, assessing whether glial activation and proinflammatory cytokine release are important contributors to various human clinical pain syndromes would appear to be warranted. This avenue of investigation is exciting in that it provides a novel target for pain control. Every therapy currently on the market explicitly targets neurons. Recognition that glial activation is a powerful driving force for exaggerated pain opens up new ways to approach effective clinical pain control.

If it is assumed, for the moment, that spinal cord proinflammatory cytokines do in fact underlie some forms of clinical pain, it is natural to inquire whether currently available drugs are potentially useful in humans. The answer is "maybe".

Presently available proinflammatory cytokine antagonists are either large proteins (IL-1 receptor antagonist; TNF binding protein), antisera (against IL-1, TNF, or IL-6), or soluble receptor/antibody hybrids.[131] None of these agents effectively cross the blood-brain barrier. Thus, unless centrally administered, none are viable options for disrupting spinal proinflammatory cytokine function in humans.

It is also possible to target proinflammatory cytokine production with other classes of drugs. For example, thalidomide and recently developed thalidomide analogs[133] inhibit the production of TNF[134], including TNF release from astrocytes[135] and microglia.[136] Such drugs have been effectively used to inhibit exaggerated pain create by peripheral TNF.[87] Inhibitors of matrix metalloproteinases, which release TNF into the extracellular space by cleaving its membrane-bound form[137], have also been used to inhibit peripheral TNF-mediated neuropathic pain.[89] However, while these drugs are effective for controlling some exaggerated pain states mediated by peripheral TNF, it is unlikely that they will be drugs of choice for control of pain at the level of the spinal cord. This is because thalidomide selectively blocks production of TNF, without effecting production of either IL-1 or IL-6.[133,138] Matrix metalloproteinase inhibitors would also fail to effect either IL-1 or IL-6 as neither of these proinflammatory cytokines is released by cleavage from the extracellular membrane, as is true for TNF. Furthermore, matrix metalloproteinases exert a wide range of effects, well beyond release of TNF. Thus, the drug actions are not sufficiently selective to be a viable option for clinical pain control.

Currently available glial metabolic inhibitors (fluorocitrate or fluoroacetate[139]) are also not appropriate for clinical consideration. These compounds are useful in animal studies when used at low doses and short post-drug testing intervals because they first act to inhibit glial metabolism (for discussion, see[42]). However, neuronal function can also be affected at higher doses and longer time intervals. Furthermore, nonselective suppression of all glial function is not a viable approach for use in the clinic.

There are two classes of compounds that may be useful to test clinically. First are inhibitors of p38 MAP kinases, recently discovered members of the MAP kinase family. As reviewed previously, we have shown that i.t. administration of a p38 MAP kinase inhibitor (CNI-1493) blocks: (a) thermal hyperalgesia after s.c. formalin[27], (b) both thermal hyperalgesia and mechanical allodynia after spinal immune activation by i.t. gp120[42], and (c) both unilateral and bilateral mechanical allodynias induced by immune activation around a single healthy nerve (sciatic inflammatory neuritis; SIN).[28-30] Furthermore, we have recently demonstrated that systemically administered CNI-1493 abolishes both thermal hyperalgesia and mechanical allodynia induced by i.t. gp120.[30,71] p38 MAP kinase is one of the major signal transduction pathways used by glia and immune cells, leading to the production of TNF, IL-1 and IL-6.[67,68] It also stimulates glial NO production.[69] Furthermore, p38 MAP kinase is the major signal transduction pathway activated upon binding of TNF, IL-1 and IL-6 to other cells.[72,73,140] Thus, p38 MAP kinase inhibitors would be predicted to block a wide variety of pain states by disruption of NO production, disruption of proinflammatory cytokine production and/or disruption of proinflammatory cytokine downstream effects. Cell function would not be globally suppressed, as production of other cytokines and cellular proteins is unaffected by p38 MAP kinase inhibitors.[141,142]

One caveat to this discussion is that few investigations have yet examined whether p38 MAP kinase inhibitors exert direct effects on neurons. What little is known is that activation of p38 MAP kinase can stimulate neuronal apoptosis during early development[143] and in response to trauma.[144] Clearly, whether p38 MAP kinase inhibitors have any detrimental effects on neuronal function needs to be explored. On the other hand, systemically administered p38 MAP kinase inhibitors have successfully passed Phase I and Phase II clinical trials for unrelated applications, so their potential safety ranges in humans are currently being established (K.J. Tracey, personal communication).

The second class of compounds that may be worth pursuing clinically is the family of xanthine derivatives that include pentoxifylline and propentofylline. These compounds cross the blood-brain barrier, which enhances their potential clinical application. These compounds have also successfully passed human clinical trials for use in non-pain conditions.[145,146] They can inhibit production of TNF, IL-1, and oxygen free radicals from glial cells[147-149] cf[150], inhibit IL-6 production[151] cf[152], suppress microglial and astrocyte activation[153], inhibit microglial proliferation[154], enhance uptake of extracellular excitatory amino acids by glia[155,156], enhance glial release of nerve growth factors[157], suppress neuronal intracellular calcium accumulation[156], increase anti-inflammatory cytokine (IL-4 and IL-10) production[158], and increase extracellular adenosine by inhibition of glial adenosine transporters.[159] The increase in extracellular adenosine is notable in that adenosine activates potassium and chloride conductances in neurons, which limits synaptically evoked depolarization, thus counteracting calcium ion influx through voltage-dependent and NMDA receptor-operated ion channels.[160] Delivered i.t., adenosine both decreases the size of the human body area that develops topical mustard oil-induced mechanical allodynia[161] and inhibits substance P and glutamate release and postsynaptic actions.[162] Taken together, pentoxifylline and propentofylline appear to have diverse actions, but all the documented actions appear to favor the desired result of relieving exaggerated pain states.

In conclusion, glial activation and its associated proinflammatory cytokine release appear to drive exaggerated pain states within the central nervous system. These effects have been best documented in spinal cord, but growing evidence points to pain modulatory effects of proinflammatory cytokines in brain as well. The fact that dramatic changes in pain are induced by these glial products suggests that developing drugs that target proinflammatory cytokine actions may provide novel approaches to the control of human clinical pain.

Acknowledgments

This work was supported by NIH Grants MH55283, NS38020, MH01558, MH45045, and MH00314.

References

1. Basbaum AI, Fields HL. Endogenous pain control systems: Brainstem spinal pathways and endorphin circuitry. Annu Rev Neurosci 1984; 7:309-338.
2. Sandkuhler J. The organization and function of endogenous antinociceptive systems. Prog Neurobiol 1996; 50:49-81.
3. Kelly DD, ed. Stress-induced analgesia. New York: NYAS, 1986.
4. Oka T, Hori T. Brain cytokines and pain. In: Watkins LR, Maier SF, eds. Cytokines and pain. Basel: Birkhäuser Verlag, 1999:183-204.
5. Watkins LR, Maier SF, eds. Cyotkines and pain. Basel: Birkhauser Verlag, 1999.
6. Watkins LR, Maier SF. The case of the missing brain: Arguments for a role of brain-to-spinal cord pathways in pain facilitation. Behav Brain Sci 1997; 20:469.
7. Cova JL, Aldskogius H. Effect of nerve section on perineuronal glial cells in the CNS of rat and cat. Anat Embryol (Berl) 1984; 169:303-307.
8. Gehrmann J, Monaco S, Kreutzberg GW. Spinal cord microglial cells and DRG satellite cells rapidly respond to transection of the rat sciatic nerve. Restorative Neurol Neurosci 1991; 2:181-198.
9. Gilmore SA, Skinner RD. Intraspinal non-neuronal cellular responses to peripheral nerve injury. Anatomical Record 1979; 194:369-388.
10. Brenner M. Structure and transcriptional regulation of the GFAP gene. Brain Path 1994; 4:245-257.
11. Garrison CJ, Dougherty PM, Kajander KC et al. Staining of glial fibrillary acidic protein (GFAP) in lumbar spinal cord increases following a sciatic nerve constriction injury. Brain Res 1991; 565:1-7.
12. Garrison CJ, Dougherty PM, Carlton SM. GFAP expression in lumbar spinal cord of naive and neuropathic rats treated with MK-801. Exp Neurol 1994; 129:237-43.
13. Meller ST, Dyskstra C, Grzybycki D et al. The possible role of glia in nociceptive processing and hyperalgesia in the spinal cord of the rat. Neuropharmacology 1994; 33:1471-1478.
14. Watkins LR, Deak T, Silbert L et al. Evidence for involvement of spinal cord glia in diverse models of hyperalgesia. Proc Soc Neurosci 1995; 21:897.
15. Sweitzer SM, Colburn RW, Rutkowski M et al. Acute peripheral inflammation induces moderate glial activation and spinal IL-1 beta expression that correlates with pain behavior in the rat. Brain Res 1999; 829:209-221.
16. Fu K-Y, Light AR, Matsushima GK et al. Microglial reactions after subcutaneous formalin injection into the rat hind paw. Brain Res 1999; 825:59-67.
17. DeLeo JA, Colburn RW. Proinflammatory cytokines and glial cells: Their role in neuropathic pain. In: Watkins L, ed. Cytokines and Pain. Basel: Birkhauser, 1999:159-182.
18. DeLeo JA, Colburn RW, Rickman AJ et al. Cytokine and growth factor immunohistochemical spinal profiles in two animal models of mononeuropathy. Brain Res 1997; 759:50-57.
19. Colburn RW, Rickman AJ, DeLeo JA. The effect of site and type of nerve injury on spinal glial activation and neuropathic pain behavior. Exp Neurol 1999; 157:289-304.

20. Coyle DE. Partial peripheral nerve injury leads to activation of astroglia and microglia which parallels the development of allodynic behavior. Glia 1998; 23:75-83.
21. Schwei MJ, Honore P, Rogers SD et al. Neurochemical and cellular reorganization of the spinal cord in a murine model of bone cancer pain. J Neurosci 1999; 19:10886-10897.
22. Hashizume H, DeLeo JA, Colburn RW et al. Spinal glial activation and cytokine expression after lumbar root injury in the rat. Spine 2000; 25:1206-1217.
23. Popovich PG, Wei P, Stokes BT. Cellular inflammatory response after spinal cord injury in Sprague-Dawley and Lewis rats. J Comp Neurol 1997; 377:443-464.
24. Wiertelak EP, Furness LE, Horan R et al. Subcutaneous formalin produces centrifugal hyperalgesia at a non-injected site via the NMDA-nitric oxide cascade. Brain Res 1994; 649:19-26.
25. Watkins LR, Wiertelak EP, Goehler L et al. Neurocircuitry of illness-induced hyperalgesia. Brain Res 1994; 639:283-299.
26. Milligan ED, O'Connor KA, Nguyen KT et al. Intrathecal HIV-1 envelope glycoprotein gp120 enhanced pain states mediated by spinal cord proinflammatory cytokines. J Neurosci 2001; 21:2808-2819.
27. Watkins LR, Martin D, Ulrich P et al. Evidence for the involvement of spinal cord glia in subcutaneous formalin induced hyperalgesia in the rat. Pain 1997; 71:225-35.
28. Chacur M, Milligan ED, Gazda LS et al. A new model of sciatic inflammatory neuritis (SIN): induction of unilateral and bilateral mechanical allodynia following acute unilateral peri-sciatic immune activation in rats. Pain 2001; in press.
29. Hammack SE, Milligan ED, Gazda L et al. Spinal interleukin-1 receptor antagonist blocks allodynia from zymosan-induced neuritis. Proc Soc Neurosci 1999; 25:1443.
30. Watkins LR, Maier SF. The pain of being sick: implications of immune-to-brain communication for understanding pain. Ann Rev Psych 2000; 1:29-57.
31. Sweitzer SM, Schubert P, DeLeo JA. Propentofylline, a glial modulating agent, exhibits anti-allodynic properties in a rat model of neuropathic pain. J Pharmacol Exp Ther 2001; 297(3):1210-1217.
32. Milligan ED, Mehmert K, Hinde JL et al. Thermal hyperalgesia and mechanical allodynia produced by intrathecal administration of the human immunodeficiency virus-1 envelope glycoprotein, gp120. Brain Res 2000; 861:105-116.
33. Murphy S, ed. Astrocytes: Pharmacology and Function. San Diego: Academic Press, 1993.
34. Kettenmann H, Ransom BR, eds. Neuroglia. New York: Oxford University Press, 1995.
35. Kreutzberg GW. Microglia: a sensor for pathological events in the CNS. Trends Neurosci 1996; 19:312-8.
36. Malcangio M, Bowery NG, Flower RJ et al. Effect of interleukin-1beta on the release of substance P from rat isolated spinal cord. Eur J Pharmacol 1996; 299:113-118.
37. Inoue A, Ikoma K, Morioka N et al. Interleukin-1beta induces substance P release from primary afferent neurons through the cyclooxygenase-2 system. J Neurochem 1999; 73:2206-2213.
38. Southall MD, Michael RL, Vasko MR. Intrathecal NSAIDS attenuate inflammation-induced neuropeptide release from rat spinal cord slices. Pain 1998; 78:39-48.
39. Aimar P, Pasti L, Carmignoto G et al. Nitric oxide-producing islet cells modulate the release of sensory neuropeptides in the rat substantia gelatinosa. J Neurosci 1998; 18:10375-10388.
40. Kommers T, Vinade L, Pereira C et al. Regulation of the phosphorylation of glial fibrillary acidic protein (GFAP) by glutamate and calcium ions in slices of immature rat spinal cord: compariosn with immature hippocampus. Neurosci Lett 1998; 248:141-143.
41. Inagaki N, Fukui H, Ito S et al. Type-2 astrocytes show intracellular Ca2+ elevation in response to various neuroactive substances. Neurosci Lett 1991; 128:257-260.
42. Milligan ED, Mehmert KK, Hinde JL et al. Thermal hyperalgesia and mechanical allodynia produced by intrathecal administration of the Human Immunodeficiency Virus-1 (HIV-1) envelope glycoprotein, gp120. Brain Res 2000; 861:105-116.
43. Baldwin SA, Broderick R, Blades DA et al. Alterations in temporal/spatial distribution of GFAP- and vimentin-positive astrocytes after spinal cord contusion with the New York University spinal cord injury device. J Neurotrauma 1998; 15:1015-1026.
44. Cho JW, Troy FA. Polysialic acid engineering: synthesis of polysialylated neoglycospingolipids by using the polysilytransferase from neuroinvasic Escherichia coli K 1. Proc Natl Acad Sci USA 1994; 91:11427-31.
45. Vartdal F, Vandivik B, Norrby E. Viral and bacterial antibody responses in multiple sclerosis. Ann Neurol 1980; 8:248-55.

46. Cunningham ET, De Souza EB. Interleukin-1 receptors in the brain and endocrine tissues. Immunol Today 1993; 14:171-176.
47. Botchkina GI, Meistrell MER, Botchkina IL et al. Expression of TNF and TNF receptors (p55 and p75) in the rat brain after focal cerebral ischemia. Mol Med 1997; 3:765-81.
48. Chambaut-Guerin AM, Rouher C, Gauthereau X. p55 tumour necrosis factor receptors distribution in neuroblastoma cells. Neuroreport 1997; 14:1451-6.
49. Ballestas ME, Benveniste EN. Interleukin-1-beta and tumor necrosis factor-alpha mediated regulation of ICAM-1 gene expression in astrocytes requires protein kinase C activity. Glia 1995; 14(4):267-78.
50. Sweitzer S, Martin D, DeLeo JA. IL-1ra and sTNFr reduces mechanical allodynia and spinal cytokine expression in a model of neuropathic pain. Neuroscience 2001; 103(2):529-539.
51. DeLeo JA, Rutkowski MD, Stalder AK et al. Transgenic expression of TNF by astrocytes increases mechanical allodynia in a mouse neuropathy model. Neuroreport 2000; 11:599-602.
52. Ignatowski TA, Covey WC, Knight PR et al. Brain-derived TNFa mediates neuropathic pain. Brain Res 1999; 841:70-77.
53. Oka T, Aou S, Hori T. Intracerebroventricular injection of interleukin-1b induces hyperalgesia in rats. Brain Res 1993; 624:61-68.
54. Oka T, Aou S, Hori T. Intracerebroventricular injection of interleukin-1b enhances nociceptive neuronal responses of the trigeminal nucleus caudalis in rats. Brain Res 1994; 656:236-244.
55. Oka T, Oka K, Hosoi M et al. The opposing effects of interleukin-1b microinjected into the preoptic hypothalamus and the ventromedial hypothalamus on nociceptive behavior in rats. Brain Res 1995; 700:271-278.
56. Watkins LR, Wiertelak EP, Goehler LE et al. Characterization of cytokine-induced hyperalgesia. Brain Res 1994; 654:15-26.
57. DeLeo JA, Colburn RW, Nichols M et al. Interleukin (IL)-6 mediated hyperalgesia/alloydnia and increased spinal IL-6 in a rat mononeuropathy model. Journal of Interferon and Cytokine Research 1996; 16:695-700.
58. Maier SF, Watkins LR. Cytokines for psychologists: implications of bi-directional immune-to-brain communication for understanding behavior, mood, and cognition. Psych Rev 1998; 105:83-107.
59. Watkins LR, Maier SF. Illness-induced hyperalgesia: mediators, mechanisms, and implications. In: Watkins LR, Maier SF, editors. Cytokines and Pain. Basel: Birkhauser, 1999:39-58.
60. Watkins LR, Maier SF, Goehler LE. Immune activation: The role of pro-inflammatory cytokines in inflammation, illness responses, and pathological pain states. Pain 1995; 63:289-301.
61. Buttini M, Limonta S, Boddeke HW. Peripheral administration of lipopolysaccharide induces activation of microglial cells in rat brain. Neurochem Int 1996; 29:25-35.
62. Koka P, He K, Zack JA et al. Human immunodeficiency virus 1 envelope proteins induce interleukin-1, tumor necrosis factor alpha, and nitric oxide in glial cultures derived from fetal, neonatal and adult human brain. J Exp Med 1995; 182:941-951.
63. Kong LY, Wilson BC, McMillian MK et al. The effects of the HIV-1 envelope protein gp120 on the production of nitric oxide and proinflammatory cytokines in mixed glial cultures. Cell Immunol 1996; 172:77-83.
64. Vesce S, Bezzi P, Rossi D et al. HIV-1 gp120 glycoprotein affects the astrocyte control of extracellular glutamate by both inhibiting the uptake and stimulating the release of the amino acid. FASEB Lett 1997; 411:107-109.
65. Ushijima H, Nishio O, Klocking R et al. Exposure to gp120 of HIV-1 induces an increased release of arachidonic acid in rat primary neuronal cell culture followed by NMDA receptor-mediated neurotoxicity. Eur J Neurosci 1995; 7:1353-1359.
66. Watkins LR, Hansen MK, Nguyen KT et al. Dynamic regulation of the proinflammatory cytokine, interleukin-1 beta: Molecular biology for non-molecular biologists. Life Sci 1999; 65:449-481.
67. Molina-Holgado F, Lledo A, Guaza C. Anandamide supresses nitric oxide and TNF-alpha responses to Theiler's virus or endotoxin in astrocytes. Neuroreport 1997; 27:1929-33.
68. Bhat NR, Zhang P, Lee JC et al. Extracellular signal-regulated kinase and p38 subgroups of mitogen-activated protein kinases regulate inducible nitric oxide synthase and tumor necrosis factor-alpha gene expression in endotoxin-stimulated primary glial cultures. J Neurosci 1998; 18:1633-41.
69. Bhat NR, Zhang P, Bhat AN. Cytokine induction of inducible nitric oxide synthase in an oligodendrocyte cell line: role of p38 mitogen-activated protein kinase activation. J Neurochem 1999; 72:474-8.

70. Lazarini F, Casanova P, Tham TN et al. Differential signalling of the chemokine receptor CXCR4 by stromal cell-derived factor 1 and the HIV glycoprotein in rat neurons and astrocytes. Eur J Neurosci 2000; 12:117-25.
71. Milligan ED, O'Connor KA, Armstrong CB et al. Systemic administration of CNI-1493, a p38 MAP kinase inhibitior, blocks HIV-1 gp 120-induced enhanced pain states in rats. Journal of Pain 2001; in press.
72. Ridley SH, Sarsfield SJ, Lee JC et al. Actions of IL-1 are selectively controlled by p38 mitogen-activated protein kinase: regulation of prostaglandin H synthase-2, metalloproteinases, and IL-6 at different levels. J Immunol 1997; 158:3165-73.
73. Ridley SH, Dean JL, Sarsfield SJ et al. A p38 MAP kinase inhibitor regulates stability of interleukin-1-induced cyclooxygenase-2 mRNA. FEBS Lett 1998; 439:75-80.
74. Holguin A, O'Connor KA, Bravmann C et al. Inhibition of nitric oxide blocks HIV-1 gp120-induced spinal proinflammatory cytokine production and low threshold mechanical allodynia. Proc Soc Neurosci 2001; 27: in press.
75. Maves TJ, Pechman PS, Gebhart GF et al. Possible chemical contribution from chromic gut sutures produces disorders of pain sensation like those seen in man. Pain 1993; 54:57-69.
76. Clatworthy AL, Hughes TK, Budelmann BU et al. Induction of a cellular defense reaction is accompanied by an increase in sensory neuron excitability in Aplysia. Journal of Neurosci 1994; 14:3263-3270.
77. Clatworthy AL, Illich PA, Castro GA et al. Role of peri-axonal inflammation in the development of thermal hyperalgesia and guarding behavior in a rat model of neuropathic pain. Neurosci Lett 1995; 184:5-8.
78. Clatworthy AL, Grose E. Immune-mediated alterations in nociceptive sensory function in Aplysia californica. J Exp Biol 1999; 202:623-630.
79. Myers RR, Wagner R, Sorkin LS. Hyperalgesic actions of cytokines on peripheral nerves. In: Watkins LR, Maier SF, eds. Cytokines and Pain. Basel: Birkause, 1999:133-158.
80. Wagner R, Myers RR. Schwann cells produce tumor necrosis factor alpha: Expression in injured and non-injured nerves. Neuroscience 1996; 73:625-629.
81. George A, Schmidt C, Weishaupt A et al. Serial determination of tumor necrosis factor-alpha content in rat sciatic nerve after chronic constriction injury. Exp Neurol 1999; 160:124-132.
82. Sommer C, Petrausch S, Lindenlaub T et al. Neutralizing antibodies to interleukin-1-receptor reduce pain associated behavior in mice with experimental neuropathy. Neurosci Lett 1999; 270:25-28.
83. Wagner R, Myers RR. Endoneurial injection of TNF alpha produces neuropathic behaviors. Neuroreport 1996; 7:2897-2901.
84. Sorkin LS, Doom CM. Epineurial application of TNF elicits an acute mechanical hyperalgesia in the awake rat. J Peripher Nerv Sys 2000; 5:96-100.
85. Eliav E, Herzberg U, Ruda MA et al. Neuropathic pain from an experimental neuritis of the rat sciatic nerve. Pain 1999; 83:169-182.
86. Clatworthy AL. Evolutionary perspectives of cytokines in pain. In: Watkins LR, Maier SF, eds. Cytokines and Pain. Basel: Birkhauser, 1999:21-38.
87. George A, Marziniak M, Schäfers M et al. Thalidomide treatment in chronic constrictive neuropathy decreases endoneurial tumor necrosis factor-alpha, increases interleukin-10 and has long-term effects on spinal cord dorsal horn met-enkephalin. Pain 2000; 88(3):267-275.
88. Lindenlaub T, Teuteberg P, Hartung T et al. Effects of neutralizing antibodies to TNF-alpha on pain-related behavior and nerve regeneration in mice with chronic constriction injury. Brain Res 2000; 866:15-22.
89. Sommer C, Schmidt C, George A et al. A metalloprotease-inhibitor reduces pain associated behavior in mice with experimental neuropathy. Neurosci Lett 1997; 237:45-48.
90. Lockwood LL, Silbert LH, Laudenslager ML et al. Anesthesia-induced modulation of in vivo antibody levels: a study of pentobarbital, chloral hydrate, methoxyflurance, halothan, and ketamine/xylazine. Anesth Analg 1993; 77:769-774.
91. Bennett GJ. Neuropathic pain. In: Wall PD, Melzack R, ed. Textbook of Pain. New York: Churchill Livingston, 1994:201-224.
92. Aloisi AM, Porro CA, Cavazzuit M et al. "Mirror pain" in the formalin test: behavioral and 2-deoxyglucose studies. Pain 1993; 55:267-273.
93. Bowsher D. Contralateral mirror-image pain following anterolateral cordotomy. Pain 1988; 33:63-65.
94. Petrovaara A. A neuronal correlate of secondary hyperalgesia in the rat spinal dorsal horn is submodality selective and facilitated by supraspinal influence. Exp Neurol 1998; 149(1):193-202.

95. Swett JE, Woolf CJ. The somatotopic organization of primary afferent terminals in the superficial laminae of the dorsal horn of the rat spinal cord. J Comp Neurol 1985; 231(1):66-77.
96. Ignatowski TA, Noble BK, Wright JR et al. Neuronal-associated tumor necrosis factor (TNFa): Its role in noradrenergic functioning and modification of its expression following antidepressant drug administration. J Neuroimmunol 1997; 79:84-90.
97. Breder CD, Dinarello CA, Saper CB. Interleukin-1 immunoreactive innervationof the human hypothalamus. Science 1988; 240:321-324.
98. Qwarnstrom EE, Page RC, Gillis S et al. Binding, internalization, and intracellular localization of interleukin-1 beta in human diploid fibroblasts. J Biol Chem 1988; 263:8261-9.
99. Mizel SB, Kilian PL, Lewis JC et al. The interleukin-1 receptor. Dynamics of interleukin-1 binding and internalization in T cells and fibroblasts. J Immunol 1987; 138:2906-2912.
100. Solari R, Smithers N, Kennard N et al. Receptor mediated endocytosis and intracellular fate of interleukin-1. Biochem Pharmacol 1994; 47:93-101.
101. Dinarello CA. Overview of inflammatory cytokines and their roles in pain. In: Watkins LR, Maier SF, eds. Cytokines and Pain. Basel: Birkhauser, 1999:1-20.
102. Wang CX, Olschowka JA, Wrathall JR. Increase of interleukin-1beta mRNA and protein in the spinal cord following experimental traumatic injury in the rat. Brain Res 1997; 759:190-96.
103. Wang CX, Nuttin B, Heremans H et al. Production of tumor necrosis factor in spinal cord following traumatic injury in rats. J Neuroimmunol 1996; 69:151-156.
104. Covey WC, Ignatowski TA, Knight PR et al. Brain-derived TNFalpha: Involvement in neuroplastic changes implicated in the conscious perception of persistent pain. Brain Res 2000; 859:113-122.
105. Bethea JR, Nagashima H, Acosta MC et al. Systemically administered interleukin-10 reduces tumor necrosis factor-alpha production and significantly improves functional recovery following traumatic spinal cord injury in rats. J Neurotrauma 1999; 16:851-863.
106. Tadano T, Namioka M, Nakagawasai O et al. Induction of nociceptive responses by intrathecal injection of interleukin-1 in mice. Life Sci 1999; 65:255-261.
107. Laughlin TM, Bethea JR, Yezierski RP et al. Cytokine involvement in dynorphin-induced allodynia. Pain 2000; 84:159-267.
108. Sweitzer SM, Arruda JL, DeLeo JA. The cytokine challenge: Methods for the detection of central cytokines in rodent models of persistent pain. In: Kruger L, ed. Methods in pain research. New York: CRC Press, 2000: in press.
109. Le YL, Shih K, Bao P et al. Cytokine chemokine expression in contused rat spinal cord. Neurochem Int 2000; 36:417-25.
110. Arruda JL, Colburn RW, Rickman AJ et al. Increase of interleukin-6 mRNA in the spinal cord following peripheral nerve injury in the rat: potential role of IL-6 in neuropathic pain. Brain Res Mol Brain Res 1998; 62:228-235.
111. Streit WJ, Semple-Rowland SL, Hurley SD et al. Cytokine mRNA profiles in contused spinal cord and axotomized facial nucleus suggest a beneficial role for inflammation and gliosis. Exp Neurol 1998; 152:74-87.
112. Yeager MP, Lunt P, Arruda J et al. Cerebrospinal fluid cytokine levels after surgery with spinal or general anesthesia. Reg Anesth Pain Med 1999; 24:557-562.
113. Ramer MS, Murphy PG, Richardson PM et al. Spinal nerve lesion-induced mechanoallodynia and adrenergic sprouting in sensory ganglia are attenuated in interleukin-6 knockout mice. Pain 1998; 78:115-121.
114. Murphy PG, Ramer MS, Borthwick L et al. Endogenous interleukin-6 contributes to hypersensitivity to cutaneous stimuli and changes in neuropeptides associated with chronic nerve constriction in mice. Eur J Neurosci 1999; 11:2243-2253.
115. Oka T, Wakugawa Y, Hosoi M et al. Intracerebroventricular injections of tumor necrosis factor-alpha induces thermal hyperalgesia in rats. Neuroimmunomodulation 1996; 3:135-140.
116. Oka T, Oka K, Hosoi M et al. Intraventricular injection of interleukin-6 induces thermal hyperalgesia in rats. Brain Res 1995; 692:123-128.
117. Yabuuchi K, Nishiyori A, Minami M et al. Biphasic effects o intracerebroventricular interleukin-1b on mechanical nociception in the rat. Eur J Pharmacol 1996; 300:59-65.
118. Bianchi M, Sacerdote P, Ricciardi-Castagnoli P et al. Central effects of tumor necrosis factor-a and interleukin-1a on nociceptive thresholds and spontaneous locomotor activity. Neurosci Lett 1992; 148:76-80.

119. Adams JU, Bussiere JL, Geller EB et al. Pyrogenic doses of intracerebroventricular interleukin-1 did not induce analgesia in the rat hot-plate or cold-water tail-flick tests. Life Sci 1993; 53:1401-1409.
120. Nakamura H, Nakanishi K, Kita A et al. Interleukin-1 induces analgesia in mice by a central action. Eur J Pharmacol 1988; 149:49-54.
121. Kita A, Imano K, Nakamura H. Involvement of corticotropin-releasing factor in the antinociception produced by interleukin-1 in mice. Eur J Pharamcol 1993;2 73:317-322.
122. Bianchi M, Panerai A. CRH and the noradrenergic system mediate the antinociceptive effect of central interleukin-1a in the rat. Brain Res Bull 1995; 36:113-117.
123. Sellami S, de Beaurepaire R. Hypothalamic and thalamic sites of action of interleukin-1b on food intake, body temperature and pain sensitivity in the rat. Brain Res 1995; 694:69-77.
124. Vitkovic L, Bockaert J, Jacque C. "Inflammatory" cytokines: neuromodulators in normal brain? J Neurochem 2000; 74:457-471.
125. Henry JL, Yashpal K, Pitcher GM et al. Evidence for tonic activation of NK-1 receptors during the second phase of the formalin test in the rat. J Neurosci 1999; 19:6588-6598.
126. Palma C, Minghetti L, Astolfi M et al. Functional characterization of substance P receptors on cultured human spinal cord astrocytes: Synergism of substance P with cytokines in inducing interleukin-6 and prostaglandin E2 production. Glia 1997; 21:183-193.
127. Cadman ED, Witte DG, Lee CM. Regulation of the release of interleukin-6 from human astrocytoma. J Neurochem 1994; 63:980-987.
128. Lieb K, Schaller H, Bauer J et al. Substance P and histamine induce interleukin-6 expression in human astrocytoma cells by a mechanism involving protein kinase C and nuclear factor-IL-6. J Neurochem 1998; 70:1577-1583.
129. Luber-Narod J, Kage R, Leeman SE. Substance P enhances the secretion of tumor necrosis factor-alpha from neuroglial cells stimulated with lipopolysaccharide. J Immunol 1994; 152:819-24.
130. Martin FC, Charles AC, Sanderson MJ et al. Substance P stimulates IL-1 production by astrocytes via intracellular calcium. Brain Res 1992; 599:13-18.
131. Sandborn WJ, Hanauer SB. Antitumor necrosis factor therapy for inflammatory bowel disease: a review of agents, pharmacology, clinical results, and safety. Inflamm Bowel Dis 1999; 5:119-133.
132. Finegold AA, Mannes AJ, Iadarola MJ. A paracrine paradigm for in vivo gene therapy in the central nervous system: Treatment of chronic pain. Human Gene Therapy 1999; 10:1251-1257.
133. Corral LG, Muller GW, Moreira AL et al. Selection of novel analogs of thalidomide with enhanced tumor necrosis factor alpha inhibitory activity. Mol Med 1996; 2:506-515.
134. Zwingenberger K, Wnendt S. Immunomodulation by thalidomide: Systematic review of the literature and of unpublished observations. J Inflamm 1995; 46:177-211.
135. Gallily R, Kipper-Galperin M, Brenner T. Mycoplasma fermentans-induced inflammatory response of astrocytes: Selective modulation by aminoguanidine, thalidomide, pentoxifylline and IL-10. Inflammation 1999; 23:495-505.
136. Peterson PK, Hu S, Sheng WS et al. Thalidomide inhibits tumor necrosis factor-alpha production by lipopolysaccharide- and lipoarabinomannan-stimulated human microglial cells. J Infect Dis 1995; 172:1137-1140.
137. Matyszak MK, H. PV. Delayed-type hypersensitivity lesions in the central nervous system are prevented by inhibitors of matrix metalloproteinases. J Neuroimmunol 1996; 69:141-149.
138. Weglicki WB, Stafford RE, Dickens BF et al. Inhibition of tumor necrosis factor-alpha by thalidomide in magnesium deficiency. Mol Cell Biochem 1993; 129:195-200.
139. Hassel B, Paulsen RE, Johnson A et al. Selective inhibition of glial cell metabolism by fluorocitrate. Brain Res 1992; 249:120-124.
140. Miyazawa K, Mori A, Miyata H et al. Regulation of interleukin-1 beta-induced interleukin-6 gene expression in human fibroblast-like synoviocytes by p 38 mitogen-activated protein kinase. J Biol Chem 1998; 273:24832-8.
141. Bianchi M, Ulrich P, Bloom O et al. An inhibitor of macrophage arginine transport and nitric oxide production (CNI-1493) prevents acute inflammation and endotoxin lethality. Mol Med 1995; 1:254-66.
142. Bianchi M, Bloom O, Raabe T et al. Suppression of proinflammatory cytokines in monocytes by a tetravalent guanylhydrazone. J Exp Med 1996; 183:927-36.
143. Oo TF, Henchcliffe C, James D et al. Expression of c-fos, c-jun, and c-jun N-terminal kinase (JNK) in a developmental model of induced apoptotic death in neurons of the substantia nigra. J Neurochem 1999; 72:557-64.

144. Nakahara S, Yone K, Sakou T et al. Induction of apoptosis signal regulating kinase 1 (ASK1) after spinal cord injury in rats: possible involvement of ASK-1-JNK and -p38pathways in neuronal apoptosis. J Neuropathol Exp Neurol 1999; 58:442-50.
145. Bath PM, Bath FJ, Asplund K. Pentoxifylline, propentofylline and pentifylline for acute ischaemic stroke. Cochrane Database of Systematic Reviews 2000; CD000162.
146. Mielke R, Moller HJ, Erkinjuntti T et al. Propentofylline in the treatment of vascular dementia and Alzheimer-type dementia: Overview of phase I and phase II clinical trials. Alzheimer Disease and Associated Disorders 1998; 12:S29-S35.
147. Banati RB, Schubert P, Rothe G et al. Modulation of intracellular formation of reactive oxygen intermediates in peritoneal macrophages and microglia/brain macrophages by propentofylline. J Cereb Blood Flow Metab 1994; 14:145-149.
148. Schubert P, Rudolphi K. Interfering with the pathologic activation of microglial cells and astrocytes in dementia. Alzheimer Disease and Associated Disorders 1998; 12:S21-S28.
149. Suzumura A, Sawada M, Makino M et al. Propentofylline inhibits production of TNFalpha and infection of LP-BM5 murine leukemia virus in glial cells. J Neurovirol 1998; 4:553-559.
150. Si Q, Nakamura Y, Ogata T et al. Differential regulation of microglial activation by propentofylline via cAMP signaling. Brain Res 1988; 812:97-104.
151. Koo DJ, Yoo P, Cioffi WG et al. Mechanism of the beneficial effects of pentoxifylline during sepsis: Maintenance of adrenomedullin responsiveness and downregulation of proinflammatory cytokines. J Surg Res 2000; 91:70-76.
152. Chao CC, Hu S, Close K et al. Cytokine release from microglia: Differential inhibition by pentoxifylline and dexamethasone. J Infect Dis 1992; 166:847-853.
153. Wu YP, McRae A, Rudolphi K et al. Propentofylline attenuates microglial reaction in the rat spinal cord induced by middle cerebral artery occlusion. Neurosci Lett 1999; 260:17-20.
154. Si QS, Nakamura Y, Schubert P et al. Adenosine and propentofylline inhibit the proliferation of cultured microglial cells. Exp Neurol 1996; 137:345-349.
155. Andine P, Rudolphi KA, Fredholm BB et al. Effect of propentofylline (HWA 285) on extracellular purines and excitatory amino acids in CA1 of rat hippocampus during transient ischaemia. Br J Pharmacol 1990:814-818.
156. DeLeo J, Toth L, Schubert P et al. Ischemia-induced neuronal cell death, calcium accumulation, and glial response in the hippocampus of the Mongolian gerbil and protection by propentofylline (HWA 285). J Cereb Blood Flow Metab 1987; 7:745-751.
157. Shinoda I, Furukawa Y, Furukawa S. Stimulation of nerve growth factor synthesis/secretion by propentofylline in cultured mouse astroglial cells. Biochem Pharmacol 1990; 39:1813-1816.
158. Fushimi T, Inoue A, Koh CS et al. The effect of pentoxifylline (PTX) on Theiler's murine encephalomyelits (TMEV)-induced demyelinating disease. Cellr Immunol 1998; 186:140-146.
159. Parkinson FE, Paterson AR, Young JD et al. Inhibitory effects of propentofylline on [3H]adenosine influx. A study of three nucleoside transport systems. Biochem Pharmacol 1993; 46:891-896.
160. Schubert P, Ogata T, Marchini C et al. Protective mechanisms of adenosine in neurons and glial cells. Ann N Y Acad Sci 1997; 825:1-10.
161. Rane K, Segerdah M, Goiny M et al. Intrathecal adenosine administration: A phase 1 clinical safety study in healthy volunteers, with additional evaluation of its influence on sensory thresholds and experimental pain. Anesthesiology 1998; 89:1108-1115.
162. Sawynok J, Sweeney MI. The role of purines in nociception. Neuroscience 1989; 32:557-569.

CHAPTER 2

Peripheral Hyperalgesic Cytokines

Fernando Q. Cunha and Sérgio H. Ferreira

Hyperalgesia and Classic Inflammatory Mediators

Primary sensory neurons (PSN) become sensitized during inflammation (hyperalgesia) and as a consequence the nociceptors are able to transduce innocuous stimuli into what is perceived as pain by man or manifested as a characteristic nociceptive behaviors by animals. Classic inflammatory mediators are endogenous substances of low molecular weight, detected in inflammatory exudates. Their pharmacological effects mimic cardinal inflammatory symptoms. The major inflammatory mediator groups are peptides, eicosanoids and biologically active amines (e.g., bradykinin, BK; substance P; prostaglandins, PG; leukotrienes; histamine; 5-hydroxytryptamine). Hyperalgesia was assumed to be a result of the combined effects of the sub-thresholds of classic inflammatory mediators.[1] However, it is currently accepted that hyperalgesia results from the action of specific hyperalgesic mediators in the nociceptors. Several inflammatory stimuli cause hyperalgesia in rodents by inducing the release of eicosanoids and sympathomimetic amines.[2-4] Two other inflammatory mediators, endothelin and platelet aggregating factor (PAF), have also been shown to produce hyperalgesia both directly[5] and indirectly[6] by means of the release of cyclooxygenase (COX) metabolites. However, their contribution to the inflammatory hyperalgesia has not been clarified yet.

Activation of the nociceptors in inflammation can be induced by mechanical, chemical or thermal stimulation. Activation and sensitization have distinct molecular mechanisms. During mechanical nociception, for example, PSN activation is an ionotropic phenomenon in which tetrodotoxin-resistant Na^+ channels are involved in the conduction of the action potential and may be functionally up-regulated during hyperalgesia. PSN sensitization is a metabotropic phenomenon that is ultimately responsible for lowering the nociceptor threshold. This sensitization involves series of processes which include the stimulation of specific receptors coupled to G proteins resulting in cyclic adenosine monophosphate (cAMP) increase, protein kinase A activation, Ca^{++} channel opening and K^+ channel shutting.[7-12] Over the last ten years, it has become increasingly clear that the release of several classic hyperalgesic mediators during acute or chronic inflammation is mediated by a cascade of cytokines produced by local or migrating cells.[13-14]

Cytokines are small proteins (typically 5-30 kDa), some of which are glycoproteins and others are synthesized as larger precursors, which are then cleaved to produce the active molecules. These proteins function as soluble mediators and are produced by numerous cell types in response to a wide variety of stimuli. Cytokines can be produced by more than one cell type, in a number of tissues and can work in an autocrine, paracrine or

Immune Mechanisms of Pain and Analgesia, edited by Halina Machelska and Christoph Stein.

endocrine manner by binding with the receptors of target cells and inducing the release of other cytokines or classic hyperalgesic mediators. Since the cytokines have a number of regulatory roles, they are rarely produced at a constant rate and their production is usually induced or suppressed by specific substances generated during the development of inflammation. Consequently their half-lives are generally short as they are often inactivated or cleared. Studies of the biological actions of cytokines, both in vitro and in vivo have yielded an immense repertoire of results. However, it is becoming apparent that a specific physio-pathological cellular environment frequently defines the release or role of the cytokines. Resident cells, in the early hyperalgesic response release cytokines while migrating cells (neutrophils, eosinophils and lymphocytes) further contribute to the intensity and duration of acute hyperalgesia.

During the early 1970s, it was assumed that classic inflammatory mediators were released as a result of direct inflammatory stimuli, either from cells or by the activation of plasma systems. When it was discovered that aspirin-like drugs inhibit PG synthesis[15-17] it was thought that PG release following the inflammatory insult was regulated by arachidonic acid via the activation of membrane phospholipase A_2. It is currently accepted that interleukin-1β (IL-1β) released by local or migrating cells stimulates the production of local PG by phospholipase A_2 and inducible COX-2.[18] A cascade of hyperalgesic cytokines, the main topic of this review precede the release of IL-1β in inflammatory hyperalgesia.[13]

Nociceptive Methods and Detection of Hyperalgesic Cytokines

Our current understanding of the role of hyperalgesic cytokines in inflammatory pain has been facilitated by the use of recombinant endotoxin-free cytokines of different species, specific cytokine monoclonal antibodies, cytokine antagonists and inhibitors of cytokine release. Antibodies have also allowed the development of specific assays for the detection of cytokines in tissues and biological fluids. The use of cytokine-receptor knock-out or enzyme-deficient animals are proving to be promising investigation tools but until now have had a limited contribution to the understanding of inflammatory hyperalgesia [19-20].

It is generally accepted that the sensitization of the PSN is the common denominator of inflammatory pain. A group of hyperalgesic cytokines seems to intermediate between initial tissue injury or recognition of non-self and the final release of classic hyperalgesic inflammatory mediators. For full characterization of the nociceptive role of a cytokine, the nociceptive behavior test employed should permit the differentiation between sensitization and activation of PSN. Thus the time interval between the first tissue treatment (the injection of an hyperalgesic mediator or inflammatory agent) and the second stimulus (mechanical, thermal or chemical) should be long when attempting to detect PSN sensitization. The second stimulus induces the standard-behavior test end point, and without sensitization of the PSN, the activating stimulus is unable to induce overt pain in man or nociceptive behavior in the animal. Despite the fact that there is no overt behavior during the time interval between both stimuli, many physiopathological and biochemical events relevant to development of inflammatory hyperalgesia occur during this period.

When the initial stimulus induces an immediate overt behavior, it is difficult to discriminate whether a nociceptive stimulus causes sensitization of the PSN, as occurs in the acetic acid-writhing and formalin tests. The pretreatment of the abdominal cavities with specific cytokine-neutralizing antibodies may demonstrate the involvement of cytokines in the nociceptive response, however, it is almost impossible to conclude whether these cytokines are sensitizing or activating the PSN. Furthermore, in this test, intraperitoneal cytokine injections may also cause stimulation of vagal afferents, [21-22] therefore adding a central component to the peripheral sensitization. These methods, though, have been extremely useful in the confirmation of the indirect mechanism of action of cytokines. Recent research has demonstrated that IL-1β acts synergically with tumor necrosis factor (TNF)-α and IL-8 in the acetic acid- and zymosan-writhing tests. These cytokines produce the nociceptive response via the release of PGs and sympathomimetic amines.[23-24]

The rat paw pressure tests have greatly contributed to the characterization of the hyperalgesic role of cytokines. In the standard rat paw pressure test an increasing pressure is applied with a thin pointed piston to a small area of the skin in the dorsal region of paw. The animal is held in a vertical or horizontal position and the behavioral end point is defined as the moment in which the paw is withdrawn from the compressing piston.[25] Our modification of the standard method employs a low constant pressure of 20 mm Hg, instead of an increasing pressure, applied with a large flat smooth piston (surface, 15 mm^2) on the dorsal surface of the hind paws of the rat.[26] The pressure is discontinued when the rat presents a typical 'freezing reaction' signaled by a brief apnea, concomitant with a retraction of the head and forepaws. PSN sensitization is usually measured by the shortening of reaction time or by the reduction of reaction intensity to the second innocuous stimulus. In the standard test of Randall and Selitto the intensity of hyperalgesia is measured by the weight (grams) applied to induce hind paw withdrawal. In our variation of the test, hyperalgesia is quantified by the measurement of the reaction time. It is plausible that these different versions of the paw pressure test detect PSN sensitization originating from different structures. The paw withdrawal response i.e., the end point of the standard Randall and Selitto method may preferentially involve the skin nociceptors, whilst our version possibly selects nociceptors present in a deeper tissue layer of the plantar region. These possibilities may explain the different sensitivities to inflammatory mediators observed with both methods. However, it should be pointed out that the type of inflammatory stimuli used has a definitive effect upon the array of cytokines released. For example, nerve growth factor (NGF) and leukotrienes play much more important roles in complete Freund's adjuvant arthritis (CFA) than in carrageenin- or bacterial endotoxin lipopolysaccharide (LPS)-induced acute hyperalgesia (see below).

Peripheral Hyperalgesia Induced by Cytokines

Five cytokines have been systematically studied to determine their hyperalgesic effect as well as their local participation in the inflammatory reactions: IL-1β, IL-1α, TNF-α, IL-6 and NGF.

Approximately ten years ago, the availability of human recombinant IL-1α and IL- 1β allowed us to make the first behavioral study on the role of IL-1 in the inflammatory response.[26] It was already known that IL-1β induces release of PG[27] and that the eicosanoids were recognized PSN sensitizers, associated with nociception.[2,7] Experiments using intraplantar (i.pl.) IL-1β injections of sub-picogram doses demonstrated IL-1β to be much more potent than IL-1α. The IL-1s were also shown to cause hyperalgesia via the release of PG, since their effects were abolished by indomethacin. Infusion of IL-1β in the isolated rabbit ear also demonstrated the potentiation of acetylcholine nociception to be parallel to the release of PG.[27] The potent hyperalgesic effect of IL-1β was later confirmed by other investigators using our version of the paw pressure test,[28] the standard Randall Selitto rat paw pressure test[29] and models of thermal hyperalgesia.[22,29-31]

In addition to IL-1β, TNF-α, IL-6 and IL-8 are generally regarded to be pro-inflammatory despite the fact that IL-6 demonstrates anti-inflammatory actions.[32-33] TNF-α, IL-6 and IL-8, when injected i.pl. in our rat paw pressure test, were also shown to possess hyperalgesic activities at doses that did not cause edema.[13,34] Qualitatively, hyperalgesic responses to IL-8 were similar to responses of IL-1α. The responses were of fast onset, with the intensity of hyperalgesia reaching a plateau within 60 min of injection and beginning to decline within six hours.[13,26,34] These results could be interpreted as an early release of prostacyclin or sympathomimetic amines, the hyperalgesic effects of which are immediate, in contrast to the delayed onset of PGE_2.[2-3] Responses to TNF-α and IL-6 were of slower onset, with the intensity of hyperalgesia reaching a plateau within 2-3 hours of injection. This response can be interpreted as resulting from an indirect process i.e., intermediate, achieved by the release of other hyperalgesic cytokines, as we showed in next experiments.[13]

We described that i.pl. high dose of IL-1β or intraperitoneal injection of IL-1β produces hyperalgesia in both hind paws. This hyperalgesia was seen to be peripheral since it was abolished by pretreatment of the contralateral paws with recombinant IL-1β-neutralizing antibodies or COX inhibitors.[26] Spinal and supraspinal sensitization may also be an important factor in the development of inflammatory hyperalgesia, particularly when the circulating levels of TNF-α and IL-1β are high. This sensitization seems to involve the stimulation of vagal afferents primarily associated with thermal hyperalgesia.[21] Thus, it is plausible that circulating IL-1β causes mainly central thermal hyperalgesia via a vagal mechanism and mechanical hyperalgesia by local peripheral action.

NGF belongs to the neurotrophin family of proteins together with brain derived-neurotrophic factor (BDNF), neurotrophin-3 (NT-3), NT-4/5 and NT-6. It is well accepted that NGF governs the innervation of target tissues during development, playing an important role in neuronal survival and maintenance of connectivity. In addition, its systemic or local administration is known to induce a long-lasting mechanical and thermal hyperalgesia in the rat.[35-38] Treatment of adult rats with a single intraperitoneal high dose of NGF results in a prolonged hypersensitivity to noxious thermal stimulation that becomes noticeable within 30 min of administration and lasts for several days. A significant mechanical hyperalgesia develops within 7 hours following injection of NGF and persists for up to 7 days.[38] In humans, intradermal recombinant human NGF induces pressure sensitization and a lowered heat-pain threshold.[39]

The Indirect Peripheral Hyperalgesic Effects of Cytokines

The use of specific receptor antagonists and inhibitors of cytokine synthesis and release have been instrumental in the understanding of the indirect action of hyperalgesic cytokines, i.e. via the release of classic inflammatory mediators. Using the rat paw pressure test and the writhing test in mice, it was shown that COX products and sympathomimetic amines are the main mediators responsible for hyperalgesia induced by carrageenin or LPS.[2-3,34,40] Experiments in which hyperalgesic responses to IL-1β, IL-6, IL-8 and TNF-α were measured subsequent to the administration of indomethacin or the β-adrenoceptor antagonist propanolol, or both, have yielded information regarding the relative contributions of COX products and sympathomimetic amines to the hyperalgesic effects of these cytokines. Indomethacin abolished the response to IL-1β and IL-6, reduced approximately by 50% the responses to TNF-α and did not affect the response to IL-8. In contrast, atenolol markedly reduced the responses to IL-8 and reduced approximately by 50% the response to TNF-α, but did not affect responses to IL-1β and IL-6. Indomethacin and atenolol, given together, abolished responses to TNF-α[13] (Figs. 1 and 2). It seems probable that IL-1β also activates phospholipase A_2 to generate eicosanoids, since the administration of arachidonic acid into the paw does not cause hyperalgesia by itself but potentiates

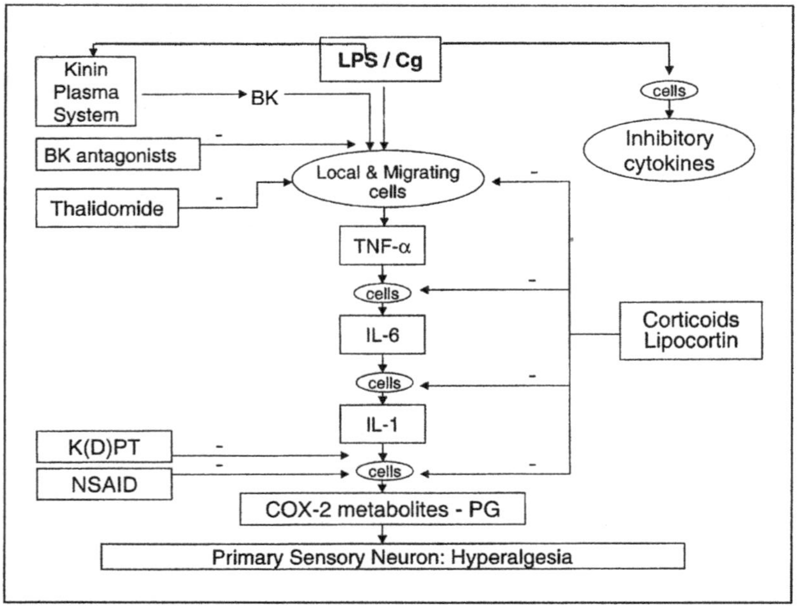

Fig. 1. Participation of hyperalgesic cytokines in the induction of cyclo-oxygenase (COX) metabolites hyperalgesia in the inflammatory response induced by gram-negative lipopolysaccharide (LPS) or carrageenin (Cg). Blockade of the cascade of hyperalgesic cytokines by bradykinin (BK) antagonists, thalidomide, or peptide Lys-D-Pro-Thr [K(D)PT] and inhibition of COX metabolites release (-) by nonsteroidal anti-inflammatory drugs (NSAID) or corticoids. TNF, tumor necrosis factor; IL, interleukin; PG, prostaglandin. See text for interpretation.

the effect of IL-1β. The hyperalgesia induced by the combined administration of arachidonic acid and IL- 1β is abolished by the pretreatment of the paws with a COX-2 inhibitor.[41]

The finding that both indomethacin and atenolol inhibited responses to TNF-α suggests a role for both IL-1β-stimulated COX products and IL-8-stimulated sympathomimetic amines in the mediation of the hyperalgesic effects of TNF-α. Based on these observations we proposed that a) IL-1β and IL-6 caused hyperalgesia by stimulating the synthesis and release of COX products, b) IL-1β induced expression of the *COX-2* gene[42] and the *phospholipase* A_2 gene[43] and IL-6 induced arachidonic acid release[32-33], and c) IL-8 caused hyperalgesia by stimulating the release of sympathomimetic amines. We pointed out earlier that the release of classic inflammatory mediators by cytokine induction depends upon the cellular environment. This can be illustrated by the fact that pretreatment of rat knee joints with the leukotriene antagonist, MK-886, inhibits the incapacitation induced by TNF-α.[44] However, MK-886 has no effect upon carrageenin induced hyperalgesia in the rat paw.

The availability of specific recombinant cytokine antisera or cytokine receptor antagonists provides us with powerful tools to investigate the release sequence of the hyperalgesic cytokine cascade during the development of inflammatory pain.[13] Hyperalgesic responses to TNF-α were inhibited (by about 50%) by IL-1β- or IL-8-neutralizing antisera and abolished by the combination of anti-IL-1β and anti-IL-8 sera. One pathway confirms the involvement of IL-1β and IL-6 in stimulating the release of COX products, whilst the other involves IL-8 which stimulates the release of sympathomimetic amines.[13,34]

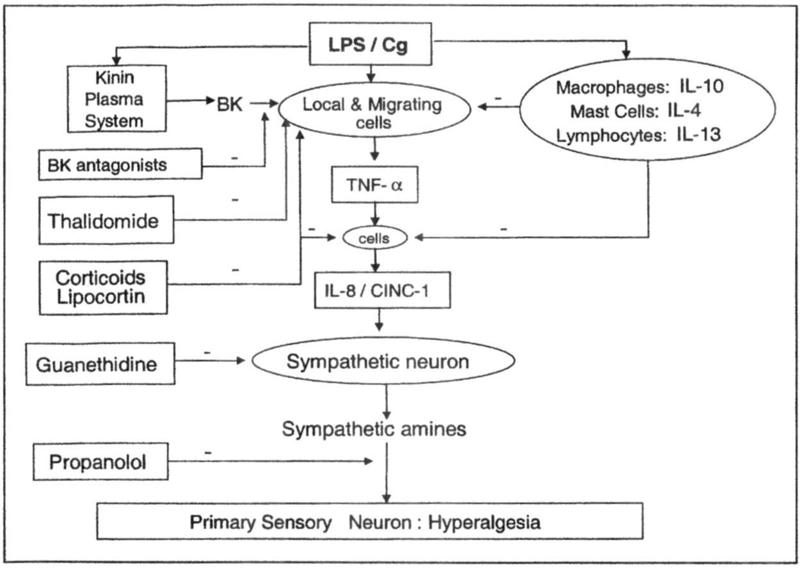

Fig. 2. Participation of hyperalgesic cytokines in the induction of sympathomimetic hyperalgesia in the inflammatory response induced by gram-negative lipopolysaccharide (LPS) or carrageenin (Cg). Blockade of sympathomimetic amines induced hyperalgesia by bradykinin (BK) antagonists, thalidomide, corticoids, guanethidine, propanolol and anti-hyperalgesic cytokines (IL-10, IL-4,and IL-13). TNF, tumor necrosis factor; IL, interleukin; CINC, cytokine-induced neutrophil chemoattractant. See text for interpretation.

Thus, in this nociceptive test the production of IL-6 and IL-8 appears to be under the control of TNF-α, rather than IL-1β. Our proposal that IL-8 mediates the release of sympathomimetic mediators was based upon the fact that the hyperalgesic effect of carrageenin, LPS, BK, TNF-α and IL-8, in the rat, was prevented by the treatment of the paws with IL-8 antisera and propanolol (Figs. 1 and 2).

In the rat, ip.l. injection of IL-1β produced an acute thermal hyperalgesia and an elevation in cutaneous NGF levels, which could be prevented by pretreatment with human recombinant IL-1 receptor antagonist. The thermal hyperalgesia but not the NGF elevation produced by intraplantar IL-1β was prevented by administration of a polyclonal anti-NGF neutralizing serum.[29] NGF, in our version of the rat paw pressure test, produces a dose-dependent hyperalgesia, which reaches a plateau within 1 hour and could be weakly inhibited by pretreatment with propanolol, indomethacin or MK866. Hyperalgesia could be inhibited (by up to 50 %) by pretreatment with three agents, indicating that NGF has either a direct hyperalgesic effect by itself or is releasing another unidentified mediator (unpublished results).

Inflammatory Stimuli and Cytokine Release

The use of monoclonal antibodies against cytokines constitutes a valuable tool for the definition of their participation in inflammatory pain. LPS- and carrageenin-induced hyperalgesia in the rat paw triggers the TNF-α-driven cascade of cytokines. Antiserum which neutralizes TNF-α, or a mixture of antisera neutralizing IL-1β and IL-8, abolish the rat paw hyperalgesic responses to carrageenin and LPS[13] as well as the knee joint incapacitation induced by carrageenin[45] and the writhings caused by intraperitoneal injection of acetic acid or zymosan.[46]

This role for TNF- α-induced IL-1β production in the early phase of the inflammatory hyperalgesia was confirmed in a study utilizing a different model of chronic and severe inflammatory hyperalgesia resulting from the i.pl. injection (i.pl.) of CFA.[47] In addition, TNF-α has been suggested to be a mediator of inflammatory pain in man, since a monoclonal antibody which neutralizes human TNF-α diminished the pain associated with rheumatoid arthritis.[48] IL-1ra significantly reduced the acute mechanical hyperalgesia produced by CFA.

Increased levels of NGF exist in CFA-induced inflammation and CFA-induced mechanical and thermal hyperalgesia can be substantially reduced by pretreatment with anti-NGF serum without reducing the elevation in IL-1β.[29] Recently we observed in the paw pressure test in which the pretreatment of the rat paws with polyclonal anti-NGF neutralizing serum (in a dose that abolishes NGF-induced hyperalgesia), had no effect upon carrageenin-induced hyperalgesia (unpublished observations). The intense tissue injury that occurs in CFA-induced hyperalgesia may account for the presence of NGF in this stimulus not seen in carrageenin-induced inflammatory hyperalgesia.

Our suggestion that IL-8 participates in the cytokine cascade was supported by the fact that antibodies to human IL-8 were able to partially (by 50%) antagonize carrageenin-induced hyperalgesia. We later realized that the rat does not produce IL-8, but instead releases cytokine-induced neutrophil chemoattractant-1 (CINC-1), IL-8 related rat chemokine (CXC subfamily). CINC-1 shares more than 80% homology with IL-8 and, therefore, the ability of IL-8 to promote hyperalgesia in rats could be explained by the fact that IL-8 and CINC-1 share the same receptor (CXCR2).[49] Preliminary results from our laboratories confirm this suggestion since pretreatment of the rat paw with antiserum to human IL-8 prevents the hyperalgesic effects of CINC-1 and antiserum to rat CINC-1 decreases carrageenin-induced hyperalgesia by 50% (Fig. 2).

Bradykinin and Cytokine Release

BK has long been regarded to be an important mediator of inflammatory pain and has a dual contribution in that it can activate or sensitize the PSNs. BK causes hyperalgesia in both behavioral and electrophysiological experimental models of inflammation.[50-53] Conversely, there is substantial research to show that inflammatory stimuli, such as carrageenin and LPS, activate the plasma kinin system.[54-55] Although BK may directly sensitize the PSN, LPS- and carrageenin-induced hyperalgesia in the rat paw triggers the TNF-α-driven cytokine cascade via the release of BK.[14,56] In fact, TNF-α-neutralizing antiserum was seen to abolish the hyperalgesic responses to carrageenin, low doses of LPS, kallidin and BK. The BK_1 and BK_2 receptor specific antagonists, however, inhibited the hyperalgesic action of kinins, carrageenin and low doses of LPS, but failed to inhibit responses to TNF-α, IL-8, IL- 1β, PGE_2 and dopamine. These observations support the idea that when BK is generated by inflammatory stimuli it may induce the release of TNF-α, thus initiating the hyperalgesic cytokine cascade. We can deduce that BK is released at the beginning of inflammation, because treatment of paws with the BK antagonists 2 hours after challenge with LPS, carrageenin or BK causes no hyperalgesia. Interestingly, BK also causes the release of TNF-α by concomitant stimulation of the BK_1 and BK_2 receptors, demonstrated by the use of a single specific antagonist of either BK_1 or BK_2 which almost abolishes BK-induced hyperalgesia.

The ability of BK to release TNF-α was supported by the in vitro observation that BK-stimulated macrophages produce both soluble and membrane-associated TNF-α.[57] The importance of kinins in the initiation of the hyperalgesic cytokine cascade in other models of inflammation is not, as yet, known. It is plausible that LPS, for example, may not be able to activate the kinin system as efficiently in other species as in the rat. In such cases LPS may directly activate TNF-α release, as observed in rats administered with intraplantar high doses of LPS.

Limitation of the Release and Action of Hyperalgesic Cytokines by Analgesic Cytokines

Of the cytokines associated with anti-inflammatory and analgesic effects the IL-4, IL-10, IL-13 and IL-1 receptor antagonist (IL-1ra) are the most thoroughly studied. Except for IL-1ra, the anti-inflammatory activities of these cytokines are the consequence of their capacities to inhibit the production and action of inflammatory cytokines, such as TNFα, IL-1β, IL-6 and chemokines in addition to their ability to inhibit COX-2 induction[58-65] (Figs. 2 and 3).

IL-4 is produced mainly by Th_2 lymphocytes and mast cells.[62,66] IL-4 suppresses the production of IL-1, TNFα, IL-8 and IFNg, whilst up-regulating the production of IL-1ra by LPS-stimulated monocytes.[67-68] Other, essentially anti-inflammatory, effects of IL-4 include inhibition of COX-2 induction.[69-70] IL-13, which is produced mainly by the Th_2 lymphocytes and by mast cells[71-72] shares a number of biological properties with IL-4, including the inhibition of the production of pro-inflammatory cytokines and eicosanoids.[58,61,63-65,70] IL-10, a product of T lymphocytes and monocytes, inhibits cytokine production by Th_1 lymphocytes[73-75] and is thought to play a role in inhibiting the delayed type hypersensitivity responses[74] and in suppressing macrophage functions, such as class II major histocompatibility complex expression,[75] adhesion[74] and the synthesis of cytokines, including IL-1, IL-6, IL-8, granulocyte colony-stimulating factor (G-CSF), granulocyte-macrophage colony-stimulating factor (GM-CSF) and TNFα.[76,74] In addition, IL-10 up-regulates the expression of IL-1ra.[77]

These cytokines, together with IL-1 receptor antagonist (IL-1ra) act as 'functional antagonists' by inhibiting the production of pro-inflammatory cytokines. The duration of the release and effect of inflammatory cytokines is curtailed by the release of anti-inflammatory cytokines and IL-1ra. These endogenous substances are released later on and persist for longer in the acute inflammatory response than the hyperalgesic cytokines.

Given the capacity of these anti-inflammatory cytokines to inhibit the production of TNFα, IL-1β, IL-6 and IL-8, we investigated their effect in the described hyperalgesic cytokine cascade. We also evaluated if their presence in the acute inflammatory response partially limited the intensity of the hyperalgesic response. In addition, the cellular source responsible for their release was also investigated.[13-14,23,26,45,56,78-81] Pretreatment (30 min before hyperalgesic cytokine challenge) of the paws with IL-4, IL-10 and IL-13 blocked in a dose-dependent manner (up to 90%) the hyperalgesia induced by carrageenin, BK and TNF-α, but did not affect responses to IL-8 and PGE_2. IL-1β hyperalgesia was inhibited by IL-10. Curiously, inhibition of the hyperalgesic effect of IL-1β occurred if pretreatment with IL-4 and IL-13 was performed twice, 2 and 12 hours before IL-1β challenge. This long time interval for the inhibition of hyperalgesia induced by IL-4 and IL-13 may be irrelevant in chronic inflammation when a continuous release is expected. The capacity of IL-10, IL-4 and IL-13 to inhibit, concomitantly, both IL-1β production and

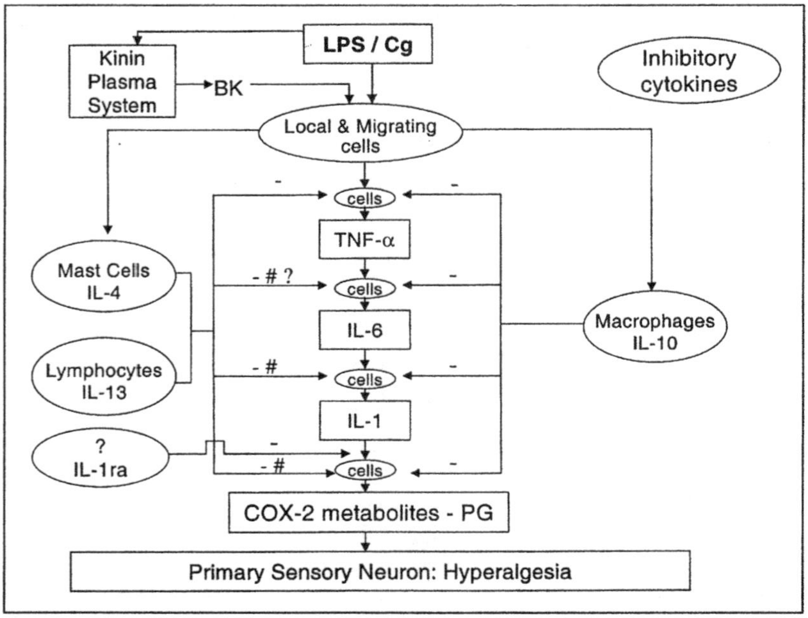

Fig. 3. Participation of hyperalgesic cytokines in the induction of cyclo-oxygenase (COX) metabolites hyperalgesia in the inflammatory response induced by gran-negative lipopolysaccharide (LPS) or carrageenin (Cg). Blockade of the cascade of hyperalgesic cytokines by the anti-hyperalgesic cytokines (interleukin (IL)-4, -10, -13 and IL-1ra). See text for interpretation.

IL-1β-stimulated PGE_2 production aids in ensuring that the cascade of mediators that cause inflammatory hyperalgesia is closely regulated. The fact that IL-10 IL-4 and IL-13 were unable to inhibit PGE2-induced hyperalgesia was expected and this, in addition to their inability to block the hyperalgesia induced by IL-8, indicates that they do not affect the release of sympathomimetic amines.

Neutralizing antibodies were used to show that endogenous IL-10, IL-4 and IL-13 play a role in limiting the development of inflammatory hyperalgesia. The antibodies potentiated the hyperalgesia triggered by carrageenin, BK and TNF-α.[78,81] The finding that the anti-IL-13 and anti-IL-4 sera did not enhance the hyperalgesic responses to IL-1β and IL-8 suggests that the main effect of endogenous IL-13 on acute inflammatory hyperalgesia is the inhibition of the release of cytokines rather than the inhibition of the production of 'downstream' mediators (PGs and sympathetic mediators) that sensitize nociceptors.[3,82] The described potentiation of hyperalgesia by anti-IL-13 and anti-IL-4 sera did not occur when athymic or mast-cells depleted rats were used, respectively. Thus, it seems that IL-4 released by mast-cells and IL-13 released by lymphocytes are limiting the intensity of inflammatory hyperalgesia.

Another inhibitory cytokine released during the inflammatory process either in animals models or in human inflammatory disease is IL-1ra.[83] IL-1ra is a specific receptor antagonist that competitively inhibits the binding of IL-1α and IL-1β to human and animal type I and II IL-1 receptors.[84-85] IL-1ra has a beneficial effect in several experimental diseases, such as sepsis, colitis, arthritis and diabetes and it is presently being tested in humans for use in the treatment of arthritis, septic shock and leukemia.[83,86-90] In chronic inflammation, IL-1ra is released concomitantly with the inflammatory cytokines and seems to contribute to the reduction of the inflammatory response.[91]

We tested in a mechanical hyperalgesia model in rats the possibility that IL-1ra limits inflammatory hyperalgesia by inhibiting the IL-1-induced eicosanoid component of nociception. In this test IL-1ra inhibited, in a dose-dependent manner, the effect of IL-1β. Local pretreatment with IL-1ra partially inhibited the hyperalgesic responses to i.pl. injection of LPS, carrageenin, BK and TNF-α, but not responses to PGE_2, IL-8 and dopamine.[79] IL-1ra, injected intraperitoneally, also inhibited the nociceptive writhing response induced by intraperitoneal injection of acetic acid. Anti IL-1ra antiserum injected i.pl. potentiates the hyperalgesic responses to the i.pl. administration of LPS, carrageenin, BK, TNF-α and IL-1β but not the response to IL-8. These results indicatë that IL-1ra is released during the inflammatory response although the cell source was identified.

The Cellular Environment and Cytokine Release

Our current understanding is moving towards the notion that the cellular environment selects the inflammatory mediators to be released and may define the physiopathological effect of the mediator. The resident macrophages seem to play an important role in the recruitment of neutrophils, a theory supported by the observation that macrophage antiserum markedly blocks the early stage of inflammatory neutrophil migration.[92] Mononuclear cells have also been seen to contribute to inflammatory nociception. Nociceptive writhing responses are increased or reduced when the peritoneal cell population is increased or diminished, respectively. [46] The importance of neutrophil recruitment for the development of hyperalgesia is demonstrated by the reduction of nociception in leucopenic animals.[47] Leukotrienes are powerful chemotactic factors in neutrophil function and, indeed, the hyperalgesia induced by leukotriene B4 has been shown to be leukocyte dependent.[4] Intraplantar injection of NGF results in local neutrophil accumulation within 3 hours of injection and this accumulation is lypoxygenase dependent. Animals in which neutrophils have been depleted do not demonstrate a thermal hyperalgesia in response to NGF, whilst prior degranulation of mast cells abolishes the early NGF-induced component of hyperalgesia.[47,93-94] The cellular environment changes throughout the duration of inflammation and, in turn, the role of the cytokine may also change. This occurrence is illustrated by the effect of TNF-α, which when administered intrarticularly into a rat knee joint primed with carrageenin[44] produces incapacitation, whilst having no effect in a normal paw. It is still not yet understood whether incapacitation occurs as the result of the release of other mediators or if TNF-α is able to activate the nociceptors already sensitised by a previous inflammation. We described above that anti-IL-13 serum potentiated responses to carrageenin, LPS, BK and TNF-α in normal rats but not in athymic rats. Thus, in a chronic immunological inflammation in which T lymphocytes are present, the release of IL-13 may contribute to limitation of the intensity of the inflammatory response.

Cytokines and Peripheral Memory of Hyperalgesia

Hyperalgesia may be classified as immediate, delayed or persistent depending upon the inflammatory mediator and the duration of its plateau.

Prostacyclin (an eicosanoid metabolite) produces an immediate sensitization, peaking at 30 min and subsiding within 1 hour of its injection, with no hyperalgesic plateau. Delayed hyperalgesia can be evoked by another eicosanoid metabolite, PGE_2, and the sympathomimetic agonist, dopamine[2-3] and has a slow onset, reaching a peak after 2-3 hours and having a plateau which lasts for 2 hours. Persistent hyperalgesia is induced by successive daily injections of hyperalgesic stimuli, such as PGE_2 or dopamine[95], which cause delayed hyperalgesia. After 6-9 daily injections, the sensitivity of the PSN does not return to its basal level but, instead, reaches a plateau. If this hyperalgesic plateau is maintained for a further 7-9 days, by continuing daily injections, sensitivity persists (in the absence of further injections) for more than 30 days. It should be noted, that there is no evidence that this persistent hyperalgesia is due to an ongoing inflammatory process triggered by the trauma of the i.pl. injections themselves. Animals were treated with indomethacin throughout the experiment and animals injected only with saline did not develop persistent hyperalgesia.[95] Persistent hyperalgesia is the result of a peripheral mechanism since the hyperalgesic state induced by either PGE_2 or dopamine was seen to be abolished by local treatment with dipyrone and N-methyl morphine (which does not cross the blood brain barrier).[96-99] After blockade of persistent hyperalgesia for one to 30 days it could be fully restored by a single mild and short-lived hyperalgesic stimulus.

Recently, the consequences of repeated i.pl. injections of the hyperalgesic cytokines, TNF-α, IL-1β and IL-8, have been addressed. The onset of the hyperalgesic plateau response induced by the cytokines was relatively late, compared with the response to PGE_2. However, if the duration of the hyperalgesic plateau induced by daily administration of cytokines persisted for more than eight days after the cessation of the injections, it remained for more than 30 days, as seen after hyperalgesic stimulation with PGE_2 or dopamine. In animals treated with indomethacin and atenolol there was no induction of persistent hyperalgesia with IL-1β and IL-8 respectively. These results indicate, as expected, that the hyperalgesic effect of IL-1β and IL-8 are mediated by the release of eicosanoids and sympathetic amines, respectively.[100]

These observations regarding persistent hyperalgesia point to an important role for PSN sensitization in the establishment of chronic pain. They also indicate the importance of using effective doses of peripherally-acting analgesics during the treatment of inflammatory states of long duration. The prevention of a long-lasting hyperalgesic state is crucial in order to avoid the development of persistent hyperalgesia. Once the persistent hyperalgesic state is established, COX inhibitors are ineffective and in such circumstances the only analgesics able to inhibit the ongoing hyperalgesia are drugs that directly block ongoing hyperalgesia such as dipyrone, diclofenac and flurbiprofen.[101-102]

Pharmacological Control of Hyperalgesic Cytokine Action

A number of clinically useful drugs and experimental substances are available which cause analgesia by the inhibition of several points or one point of the hyperalgesic cascade. Dexamethasone, for example, blocks the release of TNF-α, IL-8 and IL-1β and the induction of COX-2. The blockade of induction of synthesis of COX-2 seems to be via a newly synthesized endogenous peptide, lipocortin (LC). LC partially limited the release of cytokines in vivo and in vitro. Thalidomide and the peptide, Lys-D-Pro-Thr [K(D)PT] block the release of TNF-α and the hyperalgesic effect of IL-1β, respectively.

Dexamethasone

It has long been accepted that glucocorticoid drugs, such as dexamethasone, inhibit both the early and late changes that contribute to the inflammatory process. LC-1, a glucocorticoid-inducible protein of 37 kDa, has been identified as a potential endogenous mediator of the anti-inflammatory actions of glucocorticoids.[103] Recombinant human LC-1

(346 amino acids) and an N-terminal polypeptide, LC-$1_{1\text{-}888}$, mimic a variety of anti-inflammatory effects of the glucocorticoids.[104] More recently, an N-terminal peptide comprising of just 25 amino acids, LC-$1_{2\text{-}26}$ (LCPS1), has been shown to mimic a variety of anti-inflammatory effects of LC-1.[105] In mice, immunoneutralization of endogenous LC-1 with antiserum to LCPS1 exacerbated the acute inflammatory response to zymosan.[106]

The effects of dexamethasone, LCPS1 and an antiserum to LCPS1 upon the hyperalgesic activities of carrageenin, bradykinin, TNF-α, IL-1β, IL-6, IL-8, PGE_2 and dopamine were investigated in a model of mechanical hyperalgesia in rats. Hyperalgesic responses to i.pl. injections of carrageenin, BK, TNF-α, IL-1β and IL-6, but not IL-8, PGE_2 and dopamine, were inhibited by pretreatment with dexamethasone. Inhibition of hyperalgesic responses to injections of BK and IL-1β, but not PGE_2, were also inhibited by pretreatment with LCPS1[18] (Figs. 1 and 3).

The above data support the notion that induction of LC by dexamethasone plays a major role in the dexamethasone inhibition of inflammatory hyperalgesia evoked by carrageenin, BK and the cytokines TNF-α, IL-1β and IL-6 and provides additional evidence that the biological activity of LC resides within the LCPS1. Furthermore, the data suggest that inhibition by LCPS1 of COX-2-induced eicosanoid production also contributes to the anti-hyperalgesic effect of dexamethasone.

Thalidomide

Thalidomide, has been demonstrated to selectively inhibit TNF-α production by human monocytes stimulated with LPS or *Mycobacterium leprae* products.[107-108] Used in the therapy of erithema *nodosum leprosum*, an acute inflammatory state occurring in lepromatous leprosy, thalidomide is a particularly effective analgesic. The effects of thalidomide have been associated with the inhibition of TNF-α production.[109-110] Thalidomide reduces nociception in rats with chronic constriction injury of the sciatic nerve, an effect which is correlated to the reduction of TNF-α expression in the sciatic endoneural area.[111]

Using the rat paw hyperalgesia test, it was shown that thalidomide inhibited in a dose-dependent manner the hyperalgesia induced by carrageenin. However, when thalidomide was given 1 hour after carrageenin it had no antinociceptive effect. Thalidomide also blocked the effect of BK but had no effect upon the hyperalgesic effect of PGE_2, IL-1β, IL-8, and TNF-α. It has also been demonstrated that thalidomide stimulates the production of IL-10 in vivo.[112] The analgesic effect of thalidomide, however, was not inhibited by pretreatment of the rat paws with neutralizing IL-10 antibodies. The writhing response in mice to acetic acid and zymosan depends upon the synergic action of TNF-α, IL-1 and IL-8 and was blocked by thalidomide.[24] These results are consistent with an early effect of thalidomide in the cytokine cascade, i.e. inhibition of TNF-α release (Figs. 1 and 3).

The development of therapies for the inhibition of TNF-α production may be an important step forward in the management of pain. Drugs such as pentoxifylline, chlorpromazine, thalidomide and glucocorticoides have been shown to inhibit TNF-α production both in vitro and in vivo[107-108,112-114] and reduce pain in humans[115] and in animals.[111,116-117] Thalidomide may serve as a prototype for the manufacture of new analgesics if its side effects can be avoided. This seems to be possible since the major side effects of thalidomide are induced by its D-isomer, whilst its anti-inflammatory action is caused by the L-isomer.

Lys-D-Pro-Thr : An Anti-hyperalgesic IL-1β Analogue

The study which first described the hyperalgesic effect of IL-1β also identified a tripeptide analogue of IL-1β that inhibited hyperalgesic responses to IL-1β, but not responses to IL-1α.[26] The tripeptide, Lys-D-Pro-Thr [K(D)PT], which was derived from the partial IL-1β agonist, Lys^{193}-Pro-Thr^{195} ($IL\text{-}1^{193\text{-}195}$), failed to antagonize responses to IL-1β in other systems, e.g., in the (in vitro) EL-4 thymoma conversion assay (for IL-1β) or in the (in vivo) rabbit pyrogen test.[26] The possibility that the anti-hyperalgesic effect of K(D)PT was mediated centrally was regarded as unlikely since the peptide did not antagonize PGE_2-induced hyperalgesia and was not effective in a 'hot-plate test', in contrast to the centrally acting analgesic, morphine.[26]

In addition to inhibiting hyperalgesic responses to IL-1β, K(D)PT also inhibited the hyperalgesic response to the pro-inflammatory agent, carrageenin, but not the responses to PGE_2 .26 The maximum anti-hyperalgesic effect of K(D)PT was similar to that of the potent nonsteroidal anti-inflammatory drug, indomethacin, although in marked contrast to indomethacin, K(D)PT was not associated with gastric lesions. Although K(D)PT is plainly not an inhibitor of COX the fact that its effects were not additive to those of indomethacin, suggests that K(D)PT acts within the pathway that involves IL-1β-stimulated release of PGs.[26]

Inhibition of IL-1β-evoked mechanical hyperalgesia by K(D)PT was soon confirmed by another group using the same model[28] and, subsequently, K(D)PT was also shown to inhibit other (usually PG-dependent) responses involving IL-1β. K(D)PT reversed inhibition by IL-1β of (electrically-stimulated) long term potentiation (LTP) in the mossy fiber CA3 pathway in mouse hippocampal slice preparations[118] and reversed the inhibition of LTP in the Shaffer-CA1 synapses and perforant path-dentate gyrus synapses that resulted from incomplete cerebral ischaemia.[119] In addition, K(D)PT inhibited IL-1β-induced augmentation of the capsaicin-induced release of the calcitonin gene related peptide from capsaicin-sensitive nerves in the trachea, a PG-dependent response.[120] K(D)PT also inhibited $desArg^9BK$ (a BK analog)-induced (PG-dependent) mechanical hyperalgesia in rat knee joints, a property that it shares with IL-1ra.[121] The apparent specificity of K(D)PT as an inhibitor of (PG-dependent) IL-1β responses in nervous tissue alone is inferred by the fact that K(D)PT has no effect on (PG-dependent) IL-1β-evoked relaxation of rabbit isolated mesenteric arteries[122] (Fig. 1).

Conclusions

A cascade of inflammatory cytokine mediates the recognition of non-self or the direct injury of local cells and the induction of local inflammatory signs via the release of classic inflammatory mediators (eicosanoids, biologically active amines and biologically active peptides). The release of classic inflammatory mediators induces the sensitization of PSNs (hyperalgesia), the common denominator of inflammatory pain. The cytokine cascade is initiated by TNF-α release from local cells which, in turn, act upon local or migrating cells further releasing IL-1β (via IL-6 release) and CINC-1. These cytokines then induce the liberation of the eicosanoids and sympathomimetic amines, respectively. In the rat, carrageenin and low doses of LPS indirectly release TNF-α by activation of the kinin system. High doses of LPS directly release TNF-α. In some nociceptive models, IL-1β releases NGF from mast cells which indirectly causes hyperalgesia via leukotriene B4. Inflammatory hyperalgesia is curtailed by analgesic cytokines (IL-4, IL-10 and IL-13) and by the IL-1β receptor-antagonist released late or together with hyperalgesic cytokines in the acute inflammatory response. The release of cytokines depends upon the cellular environment and may define their physiopathological effect. Various clinically useful drugs and experimental substances exist which cause analgesia by inhibiting the hyperalgesic cytokine cascade. Dexamethasone blocks the release of TNF-α, IL-8, IL-6 and IL-1β and dexamethasone also indirectly blocks the induction of COX-2 by IL-1β via induction of a peptide, LC. Thalidomide blocks the release of TNF- α, whilst the synthetic peptide, K(D)PT, is orally active and inhibits the hyperalgesic effect induced by IL- β. In conclusion, the elucidation of the role of hyperalgesic cytokines in the development of inflammatory pain has allowed a fuller understanding of the analgesic mechanism of old drugs and points to new targets for the development of peripherally-acting analgesics.

References

1. Lynn B. In: Wall P, Melzack R, eds. Pain. Edinburgh: Churchill Livingstone, 1984:19-33.
2. Ferreira SH, Nakamura M, Abreu Castro MS. The hyperalgesic effects of prostacyclin and prostaglandin E2. Prostaglandins 1978; 16:31-37.
3. Nakamura M, Ferreira SH. A peripheral sympathetic component in inflammatory hyperalgesia. Eur J Pharmacol 1987; 135:145-153.
4. Levine JD, Lau W, Kwiat G et al. Leukotriene B4 produces hyperalgesia that is dependent on polymorphonuclear leukocytes. Science 1984; 225:743-745.
5. Ferreira SH, Romitelli M, de Nucci G. Endothelin-1 participation in overt and inflammatory pain. J Cardiovasc Pharmacol 1989; 13 Suppl 5:S220-S222.
6. Vargaftig BB, Ferreira SH. Blockade of the inflammatory effects of platelet-activating factor by cyclo-oxygenase inhibitors. Braz J Med Biol Res 1981; 14:187-189.
7. Ferreira SH, Nakamura M. I—Prostaglandin hyperalgesia, a cAMP/Ca2+ dependent process. Prostaglandins 1979; 18:179-190.
8. Cunha FQ, Teixeira MM, Ferreira SH. Pharmacological modulation of sencondary mediator systems—cyclic AMP and cyclic GMP—on inflammatory hyperalgesia. British Journal of Pharmacology 1999; 127:671-678.
9. Lynn B, O'Shea NR. Inhibition of forskolin-induced sensitisation of frog skin nociceptors by the cyclic AMP-dependent protein kinase A antagonist H-89. Brain Res 1998; 780:360-362.
10. Taiwo YO, Bjerknes LK, Goetzl EJ et al. Mediation of primary afferent peripheral hyperalgesia by the cAMP second messenger system. Neuroscience 1989; 32:577-580.
11. Taiwo YO, Levine JD. Further confirmation of the role of adenyl cyclase and of cAMP- dependent protein kinase in primary afferent hyperalgesia. Neuroscience 1991; 44:131-135.
12. Soares AC, Leite R, Tatsuo MA et al. Activation of ATP-sensitive K(+) channels: mechanism of peripheral antinociceptive action of the nitric oxide donor, sodium nitroprusside. Eur J Pharmacol 2000; 400:67-71.
13. Cunha FQ, Poole S, Lorenzetti BB et al. The pivotal role of tumour necrosis factor alpha in the development of inflammatory hyperalgesia. Br J Pharmacol 1992; 107:660-664.
14. Ferreira SH, Lorenzetti BB, Cunha FQ et al. Bradykinin release of TNF-alpha plays a key role in the development of inflammatory hyperalgesia. Agents Actions 1993; 38 Spec No:C7-C9.
15. Vane JR. Inhibition of prostaglandin synthesis as a mechanism of action of aspirin-like drugs. Nature 1971; 231:232-235.
16. Ferreira SH, Moncada S, Vane JR. Indomethacin and aspirin abolish prostaglandin release from the spleen. Nat New Biol 1971; 231:237-239.
17. Smith JB, Willis AL. Aspirin selectively inhibits prostaglandin production in human platelets. Nature New Biol 1971; 231:236-237.
18. Ferreira SH, Cunha FQ, Lorenzetti BB et al. Role of lipocortin-1 in the anti-hyperalgesic actions of dexamethasone. Br J Pharmacol 1997; 121:883-888.
19. Ballou LR, Botting RM, Goorha S et al. Nociception in cyclooxygenase isozyme-deficient mice. Proc Natl Acad Sci U S A 2000; 97:10272-10276.
20. Murata T, Ushikubi F, Matsuoka T et al. Altered pain perception and inflammatory response in mice lacking prostacyclin receptor. Nature 1997; 388:678-682.
21. Maier SF, Goehler LE, Fleshner M et al. The role of the vagus nerve in cytokine-to-brain communication. Ann N Y Acad Sci 1998; 840:289-300.

22. Watkins LR, Wiertelak EP, Goehler LE et al. Characterization of cytokine-induced hyperalgesia. Brain Res 1994; 654:15-26.
23. Thomazzi SM, Ribeiro RA, Campos DI et al. Tumor necrosis factor, interleukin-1 and interleukin-8 mediate the nociceptive activity of the supernatant of LPS-stimulated macrophages. Mediators of Inflammation 1997; 6:195-200.
24. Ribeiro RA, Vale ML, Ferreira SH et al. Analgesic effect of thalidomide on inflammatory pain. Eur J Pharmacol 2000; 391:97-103.
25. Randall LO, Selitto JJ. A method fo measurement of analgesic activity on inflamed tissue. Arch Intern Pharmacodyn 1957; 111:409.
26. Ferreira SH, Lorenzetti BB, Bristow AF et al. Interleukin-1 beta as a potent hyperalgesic agent antagonized by a tripeptide analogue. Nature 1988; 334:698-700.
27. Schweizer A, Feige U, Fontana A et al. Interleukin-1 enhances pain reflexes. Mediation through increased prostaglandin E2 levels. Agents Actions 1988; 25:246-251.
28. Follenfant RL, Nakamura-Craig M, Henderson B et al. Inhibition by neuropeptides of interleukin-1 beta-induced, prostaglandin-independent hyperalgesia. Br J Pharmacol 1989; 98:41-43.
29. Safieh-Garabedian B, Poole S, Allchorne A et al. Contribution of interleukin-1 beta to the inflammation-induced increase in nerve growth factor levels and inflammatory hyperalgesia. Br J Pharmacol 1995; 115:1265-1275.
30. Perkins MN, Kelly D, Davis AJ. Bradykinin B1 and B2 receptor mechanisms and cytokine-induced hyperalgesia in the rat. Can J Physiol Pharmacol 1995; 73: 832-836.
31. Watkins LR, Goehler LE, Relton J et al. Mechanisms of tumor necrosis factor-alpha (TNF-alpha) hyperalgesia. Brain Res 1995; 692: 244-250.
32. Dinarello CA. In: August JT, Anders MW, Murad F, eds. Advances in Pharmacology. London: Academic Press, 1994:21-51.
33. Dinarello CA. Proinflammatory and anti-inflammatory cytokines as mediators in the pathogenesis of septic shock. Chest 1997; 112:321S-329S.
34. Cunha FQ, Lorenzetti BB, Poole S et al. Interleukin-8 as a mediator of sympathetic pain. Br J Pharmacol 1991; 104:765-767.
35. Amann R, Schuligoi R, Herzeg G et al. Intraplantar injection of nerve growth factor into the rat hind paw: local edema and effects on thermal nociceptive threshold. Pain 1996; 64:323-329.
36. Andreev NY, Dimitrieva N, Koltzenburg M et al. Peripheral administration of nerve growth factor in the adult rat produces a thermal hyperalgesia that requires the presence of sympathetic post-ganglionic neurones. Pain 1995; 63:109-115.
37. Lewin GR, Ritter AM, Mendell LM. Nerve growth factor-induced hyperalgesia in the neonatal and adult rat. J Neurosci 1993; 13:2136-2148.
38. Rueff A, Dawson AJ, Mendell LM. Characteristics of nerve growth factor induced hyperalgesia in adult rats: dependence on enhanced bradykinin-1 receptor activity but not neurokinin-1 receptor activation. Pain 1996; 66:359-372.
39. Dyck PJ, Peroutka S, Rask C et al. Intradermal recombinant human nerve growth factor induces pressure allodynia and lowered heat-pain threshold in humans. Neurology 1997; 48:501-505.
40. Duarte ID, Nakamura M, Ferreira SH. Participation of the sympathetic system in acetic acid-induced writhing in mice. Braz J Med Biol Res 1988; 21:341-343.
41. Ferreira SH, Cunha FQ, Hyslop S. In: Willoughby DA, Tomlinson A, eds. Inducible Enzymes in the Inflammatory Response. Basel—Switzerland: Birkäuser Verlag, 1999:149-167.
42. Geng Y, Blanco FJ, Cornelisson M et al. Regulation of cyclooxygenase-2 expression in normal human articular chondrocytes. J Immunol 1995; 155:796-801.
43. Burch RM, Connor JR, Axelrod J. Interleukin-1 amplifies receptor-mediated activation of phospholipase A2 in 3T3 fibroblasts. Proc Natl Acad Sci U S A 1993; 85:6306-6309.
44. Tonussi CR, Ferreira SH. Rat knee-joint carrageenin incapacitation test: an objective screen for central and peripheral analgesics. Pain 1992; 48:421-427.
45. Tonussi CR, Ferreira SH. Tumour necrosis factor-α mediates carrageenin-induced knee-joint incapacitation and also triggers overt nociception in previously inflamed rat knee-joints. Pain 1999; 82:81-87.
46. Ribeiro RA, Vale ML, Thomazzi SM et al. Involvement of resident macrophages and mast cells in the writhing nociceptive response induced by zymosan and acetic acid in mice. Eur J Pharmacol 2000; 387:111-118.
47. Woolf CJ, Allchorne A, Safieh-Garabedian B et al. Cytokines, nerve growth factor and inflammatory hyperalgesia: the contribution of tumour necrosis factor alpha. Br J Pharmacol 1997;121: 417-424.

48. Rankin EC, Choy EH, Kassimos D et al. The therapeutic effects of an engineered human anti-tumour necrosis factor alpha antibody (CDP571) in rheumatoid arthritis. Br J Rheumatol 1995; 34:334-342.
49. Shibata F, Konishi K, Nakagawa H. Identification of a common receptor for three types of rat cytokine- induced neutrophil chemoattractants (cincs). Cytokine 2000; 12:1368-1373.
50. Dray A, Bettaney J, Forster P et al. Activation of a bradykinin receptor in peripheral nerve and spinal cord in the neonatal rat in vitro. Br J Pharmacol 1988; 95:1008-1010.
51. Handwerker HO, Reeh PW. In: Bond MR, Charlton JE, Woolf CJ, eds. Proceedings of the VIth World Congress on Pain. Amsterdan: Elsevier Science, 1991:59-70.
52. Lang E, Novak A, Reeh PW et al. Chemosensitivity of fine afferents from rat skin in vitro. Neurophysiol 1990; 63:887-901.
53. Poole S, Cunha FQ, Ferreira SH. In: Saadé NE, Apkarian AV, Jabbur SJ, eds. Pain and Neuroimmune Interactions. New York: Kluwer Academic/ Plenum Publishers, 2000:31-54.
54. Rothschild AM, Gascon LA. Sulphuric esters of polysaccharides as activators of a bradykinin- forming system in plasma. Nature 1966; 212:1364.
55. Damas J, Remacle-Volon G. Influence of a long-acting bradykinin antagonist, Hoe 140, on some acute inflammatory reactions in the rat. Eur J Pharmacol 1992; 211:81-86.
56. Poole S, Lorenzetti BB, Cunha JM et al. Bradykinin B1 and B2 receptors, tumour necrosis factor alpha and inflammatory hyperalgesia. Br J Pharmacol 1999; 126:649-656.
57. Crosara-Alberto DP, Cunha FQ, Ferreira SH. In: Faist E, ed. 4th International Congress on The Immune Consequences of Trauma, Shock and Sepsis. Bologna (Italy): Monduzzi Editore, 1997:333-337.
58. Callard RE, Matthews DJ, Hibbert L. IL-4 and IL-13 receptors: are they one and the same? Immunol Today 1996; 17:108-110.
59. Cassatella MA, Meda L, Bonora S et al. Interleukin 10 (IL-10) inhibits the release of proinflammatory cytokines from human polymorphonuclear leukocytes. Evidence for an autocrine role of tumor necrosis factor and IL-1 beta in mediating the production of IL-8 triggered by lipopolysaccharide. J Exp Med 1993; 178:2207-2211.
60. Elias JA, Lentz V, Cummings PJ. Transforming growth factor-beta regulation of IL-6 production by unstimulated and IL-1-stimulated human fibroblasts [published erratum appears in J Immunol 1991 Aug 15; 147(4):1460]. J Immunol 1991; 146:3437-3443.
61. Hart PH, Vitti GF, Burgess DR et al. Potential antiinflammatory effects of interleukin 4: suppression of human monocyte tumor necrosis factor alpha, interleukin-1, and prostaglandin E2. Proc Natl Acad Sci U S A 1989; 86:3803-3807.
62. Lord CJ, Lamb JR. TH2 cells in allergic inflammation: a target of immunotherapy. Clin Exp Allergy 1996; 26:756-765.
63. Standiford TJ, Strieter RM, Chensue SW et al. IL-4 inhibits the expression of IL-8 from stimulated human monocytes. J Immunol 1990; 145:1435-1439.
64. Deleuran B, Iversen L, Deleuran M et al. Interleukin 13 suppresses cytokine production and stimulates the production of 15-HETE in PBMC. A comparison between IL-4 and IL-13. Cytokine 1995; 7:319-324.
65. Muchamuel T, Menon S, Pisacane P et al. IL-13 protects mice from lipopolysaccharide-induced lethal endotoxemia: correlation with down-modulation of TNF-alpha, IFN-gamma, and IL-12 production. J Immunol 1997; 158:2898-2903.
66. Tuno de Lara JM, Okayama Y, Mceuen AR et al. Release and inactivation of interleukin-4 by mast cells. Chignard M, Pretolani M, Renesto P et al. Cells and cytokines in lung inflammation. Ann NY Acad Sci 1994; 725:50-58.
67. Fenton MJ, Buras JA, Donnelly RP. IL-4 reciprocally regulates IL-1 and IL-1 receptor antagonist expression in human monocytes. J Immunol 1992; 149:1283-1288.
68. Vannier E, Miller LC, Dinarello CA. Coordinated antiinflammatory effects of interleukin 4: interleukin 4 suppresses interleukin-1 production but up-regulates gene expression and synthesis of interleukin-1 receptor antagonist. Proc Natl Acad Sci U S A 1992; 89:4076-4080.
69. Seitz M, Loetscher P, Dewald B et al. Production of interleukin-1 receptor antagonist, inflammatory chemotactic proteins, and prostaglandin E by rheumatoid and osteoarthritic synoviocytes—regulation by IFN-gamma and IL-4. J Immunol 1994; 152:2060-2065.
70. Endo T, Ogushi F, Sone S. LPS-dependent cyclooxygenase-2 induction in human monocytes is down- regulated by IL-13, but not by IFN-gamma. J Immunol 1996;156:2240-2246.

71. Burd PR, Thompson WC, Max EE et al. Activated mast cells produce interleukin 13. J Exp Med 1995; 181:1373-1380.
72. McKenzie AN, Culpepper JA, de Waal MR et al. Interleukin 13, a T-cell-derived cytokine that regulates human monocyte and B-cell function. Proc Natl Acad Sci U S A 1993; 90:3735-3739.
73. Fiorentino DF, Bond MW, Mosmann TR. Two types of mouse T helper cell. IV. Th2 clones secrete a factor that inhibits cytokine production by Th1 clones. J Exp Med 1989; 170:2081-2095.
74. Fiorentino DF, Zlotnik A, Mosmann TR et al. IL-10 inhibits cytokine production by activated macrophages. J Immunol 1991; 147:3815-3822.
75. de Waal MR, Haanen J, Spits H et al. Interleukin 10 (IL-10) and viral IL-10 strongly reduce antigen-specific human T cell proliferation by diminishing the antigen-presenting capacity of monocytes via downregulation of class II major histocompatibility complex expression. J Exp Med 1991; 174:915-924.
76. de Waal MR, Abrams J, Bennett B et al. Interleukin 10(IL-10) inhibits cytokine synthesis by human monocytes: an autoregulatory role of IL-10 produced by monocytes. J Exp Med 1991; 174:1209-1220.
77. Bogdan C, Vodovotz Y, Nathan C. Macrophage deactivation by interleukin 10. J Exp Med 1991; 174:1549-1555.
78. Cunha FQ, Poole S, Lorenzetti BB et al. Cytokine-mediated inflammatory hyperalgesia limited by interleukin-4. Br J Pharmacol 1999; 126:45-50.
79. Cunha JM, Cunha FQ, Poole S et al. Cytokine-mediated inflammatory hyperalgesia limited by interleukin-1 receptor antagonist. Br J Pharmacol 2000; 130:1418-1424.
80. Ferreira SH, Lorenzetti BB, Poole S. Bradykinin initiates cytokine-mediated inflammatory hyperalgesia. Br J Pharmacol 1993; 110:1227-1231.
81. Poole S, Cunha FQ, Selkirk S et al. Cytokine-mediated inflammatory hyperalgesia limited by interleukin-10. Br J Pharmacol 1995; 115:684-688.
82. Ferreira SH, Lorenzetti BB, Correa FM. Central and peripheral antialgesic action of aspirin-like drugs. Eur J Pharmacol 1978; 53:39-48.
83. Dinarello CA, Thompson RC. Blocking IL-1: interleukin-1 receptor antagonist in vivo and in vitro. Immunology Today 1991; 12:404-410.
84. Eisenberg SP, Evans RJ, Arend WP et al. Primary structure and functional expression from complementary DNA of a human interleukin-1 receptor antagonist. Nature 1990; 343: 341-346.
85. Granowitz EV, Clark BD, Mancilla J et al. Interleukin-1 receptor antagonist competitively inhibits the binding of interleukin-1 to the type II interleukin-1 receptor. J Biol Chem 1991; 266:14147-14150.
86. Cominelli F, Nast CC, Clark BD et al. Interleukin-1 (IL-1) gene expression, synthesis, and effect of specific IL-1 receptor blockade in rabbit immune complex colitis. J Clin Invest 1990; 86:972-980.
87. Ohlsson K, Bjork P, Bergenfeldt M et al. Interleukin-1 receptor antagonist reduces mortality from endotoxin shock. Nature 1990; 348:550-552.
88. Rambaldi A, Terao M, Bettoni S et al. Expression of leukocyte alkaline phosphatase gene in normal and leukemic cells: regulation of the transcript by granulocyte colony- stimulating factor. Blood 1990; 76:2565-2571.
89. Schwab JH, Anderle SK, Brown RR et al. Pro- and anti-inflammatory roles of interleukin-1 in recurrence of bacterial cell wall-induced arthritis in rats. Infect Immun 1991; 59:4436-4442.
90. Wakabayashi Y, Maeda T, Kwok YN. Increase of p75 immunoreactivity in rat urinary bladder following inflammation. Neuroreport 1996; 7:1141-1144.
91. Ishihara Y, Nishihara T, Kuroyanagi T et al. Gingival crevicular interleukin-1 and interleukin-1 receptor antagonist levels in periodontally healthy and diseased sites. J Periodontal Res 1997; 32:524-529.
92. Ferreira SH, Souza GEP, Santos RR. In: Willoughby DA, Giroud JP, eds. Inflammation: Mechanisms and Treatment. New York: MTP, 1980:789-793.
93. Bennett G, al Rashed S, Hoult JR et al. Nerve growth factor induced hyperalgesia in the rat hind paw is dependent on circulating neutrophils. Pain 1998; 77:315-322.
94. Woolf CJ, Ma QP, Allchorne A et al. Peripheral cell types contributing to the hyperalgesic action of nerve growth factor in inflammation. J Neurosci 1996; 16:2716-2723.
95. Ferreira SH, Lorenzetti BB, De Campos DI. Induction, blockade and restoration of a persistent hypersensitive state. Pain 1990; 42:365-371.
96. Ferreira SH, Lorenzetti BB, Rae GA. Is methylnalorphinium the prototype of an ideal peripheral analgesic? Eur J Pharmacol 1984; 99:23-29.

97. Lorenzetti BB, Ferreira SH. Mode of analgesic action of dipyrone: direct antagonism of inflammatory hyperalgesia. Eur J Pharmacol 1985; 114:375-381.
98. Tonussi CR, Ferreira SH. Mechanism of diclofenac analgesia: direct blockade of inflammatory sensitization. Eur J Pharmacol 1994; 251:173-179.
99. Lorenzetti BB, Ferreira SH. The analgesic effect of quaternary analogues of morphine and nalorphine. Braz J Med Biol Res 1982; 15:285-290.
100. Ferreira SH, Sachs D, Cunha FQ et al. In: Saadé NE, Apkarian AV, Jabbur SJ, eds. Pain and Neuroimmune Interactions. New York: Kluwer Academic/Plenum Publishers, 2000:3-8.
101. Ferreira SH. In: Moncada S, Feelisch M, Busse R et al, eds. The Biology of Nitric Oxide. 8th ed. London: Portland Press, 1994:324-334.
102. Geisslinger G, Ferreira SH, Menzel S et al. Antinociceptive actions of R(-)-flurbiprofen—a noncyclooxygenase inhibiting 2-arylpropionic acid—in rats. Life Sci 1994; 54:L173-L177.
103. Flower RJ, Rothwell NJ. Lipocortin-1: cellular mechanisms and clinical relevance [see comments]. Trends Pharmacol Sci 1994; 15:71-76.
104. Relton JK, Strijbos PJ, O'Shaughnessy CT et al. Lipocortin-1 is an endogenous inhibitor of ischemic damage in the rat brain. J Exp Med 1991; 174:305-310.
105. Perretti M, Ahluwalia A, Harris JG et al. Lipocortin-1 fragments inhibit neutrophil accumulation and neutrophil- dependent edema in the mouse. A qualitative comparison with an anti- CD11b monoclonal antibody. J Immunol 1993; 151:4306-4314.
106. Perretti M, Ahluwalia A, Harris JG et al. Acute inflammatory response in the mouse: exacerbation by immunoneutralization of lipocortin 1. Br J Pharmacol 1996; 117:1145-1154.
107. Moreira AL, Sampaio EP, Zmuidzinas A et al. Thalidomide exerts its inhibitory action on tumor necrosis factor alpha by enhancing mRNA degradation. J Exp Med 1993; 177:1675-1680.
108. Sampaio EP, Sarno EN, Galilly R et al. Thalidomide selectively inhibits tumor necrosis factor alpha production by stimulated human monocytes. J Exp Med 1991; 173:699-703.
109. Mohr M. Thalidomide in leprosy therapy. Int J Other Mycobatc Dis 1971; 39:598-599.
110. Sarno EN, Grau GE, Vieira LM et al. Serum levels of tumour necrosis factor-alpha and interleukin-1 beta during leprosy reactional states. Clin Exp Immunol 1991; 84:103-108.
111. Sommer C, Marziniak M, Myers RR. The effect of thalidomide treatment on vascular pathology and hyperalgesia caused by chronic constriction injury of rat nerve. Pain 1998; 74:83-91.
112. Moreira AL, Wang J, Sarno EN et al. Thalidomide protects mice against LPS-induced shock. Braz J Med Biol Res 1997; 30:1199-1207.
113. Aarestrup FM, Goncalves-da-Costa SC, Sarno EN. The effect of thalidomide on BCG-induced granulomas in mice. Braz J Med Biol Res 1995; 28:1069-1076.
114. Ohtsuka H, Higuchi T, Matsuzawa H et al. Inhibitory effect on LPS-induced tumor necrosis factor in calves treated with chlorpromazine or pentoxifylline. J Vet Med Sci 1997; 59:1075-1077.
115. Dubost JJ, Soubrier M, Ristori JM et al. An open study of the anti-TNF alpha agent pentoxifylline in the treatment of rheumatoid arthritis. Rev Rhum Engl Ed 1997; 64:789-793.
116. Gorizontova MP, Mironova IV. [The effect of prophylactic administration of pentoxifylline (trental) on development of a neuropathic pain syndrome and microcirculatory disorders caused by it]. Biull Eksp Biol Med 1995; 119:485-487.
117. Weinberg JB, Mason SN, Wortham TS. Inhibition of tumor necrosis factor-alpha (TNF-alpha) and interleukin-1 beta (IL-1 beta) messenger RNA (mRNA) expression in HL-60 leukemia cells by pentoxifylline and dexamethasone: dissociation of acivicin- induced TNF-alpha and IL-1 beta mRNA expression from acivicin-induced monocytoid differentiation. Blood 1992; 79:3337-3343.
118. Katsuki H, Nakai S, Hirai Y et al. Interleukin-1 beta inhibits long-term potentiation in the CA3 region of mouse hippocampal slices. Eur J Pharmacol 1990; 181:323-326.
119. Yoshioka M, Itoh Y, Mori K et al. Effects of an interleukin-1beta analogue [Lys-D-Pro-Thr], on incomplete cerebral ischemia-induced inhibition of long-term potentiation in rat hippocampal neurons in vivo. Neurosci Lett 1999; 261:171-174.
120. Hua XY, Chen P, Fox A et al. Involvement of cytokines in lipopolysaccharide-induced facilitation of CGRP release from capsaicin-sensitive nerves in the trachea: studies with interleukin-1beta and tumor necrosis factor-alpha. J Neurosci 1996; 16:4742-4748.
121. Davis AJ, Perkins MN. desArg^{9}BK-induced mechanical hyperalgesia and analgesia in the rat: involvement of IL-1, prostaglandins and peripheral opioids. Br J Pharmacol (Proceedings) 1996;(Suppl. Dec.):74P.
122. Marceau F, Petitclerc E, DeBlois D et al. Human interleukin-1 induces a rapid relaxation of the rabbit isolated mesenteric artery. Br J Pharmacol 1991; 103:1367-1372.

CHAPTER 3

Cytokines and Peripheral Analgesia

Michael Schäfer

Introduction

Acute transient pain serves as a physiological warning to guard the integrity of the organism. An immediate reflex, e.g., withdrawal of a body part from a heat source, prevents tissue damage. If tissue damage occurs, an inflammatory response develops that triggers mechanisms in both the nervous and the immune systems.[1] This results in an ongoing painful state. The inflammatory response consists of a release of cell products such as protones, radicals and adenosine triphosphate, the generation of prostaglandins and bradykinin, and the secretion of cytokines such as interleukin (IL)-1, IL-6, tumor necrosis factor (TNF)-α from inflammatory cells. These inflammatory mediators evoke activation of specific ion channels through the excitation of peripheral nociceptive neurons.[1] In addition, they reduce the threshold of peripheral nociceptive neurons through activation of intracellular kinases, resulting in peripheral sensitization.[1] This interaction between the immune and nervous systems leads to an increased sensitivity to painful stimuli, i.e., hyperalgesia, and serves the protection of the injured body part to prevent further tissue damage. While this is advantageous in the early period of inflammation, it may have deleterious consequences in an advanced period of inflammation. Consistently, cytokine plasma and tissue concentrations commonly increase to a peak within the first 6 hours and return to baseline values at about 12 hours.[2] Pain as a result of neuro-immune interactions is a topic of the chapters by Watkins, and Cunha and Ferreira in this volume.

Concurrently, however, counteractive endogenous mechanisms are being established to inhibit inflammatory pain at the site of tissue injury. These mechanisms are also based on interactions between the immune and nervous systems. Primary sensory neurons express mRNA specific for μ-, δ- and κ-opioid receptors indicating their synthesis in dorsal root ganglia.[3,4] After synthesis the receptor proteins are transported from the dorsal root ganglia along the axon to the peripheral nerve terminals.[5,6] This axonal transport is directed towards the sensory nerve endings within painful inflamed tissue and is enhanced under inflammatory conditions.[5,6] In parallel, the local inflammatory process triggers an enhanced expression of endogenous ligands of these receptors, the opioid peptides, within inflammatory cells.[6] These contain mainly β-endorphin but also met-enkephalin and dynorphin.[7] The opioid peptide-containing immune cells migrate in a site-directed manner from the circulation to the painful inflamed tissue (see refs. 8, 9 and the chapters by Mousa, and Machelska in this volume). This migration is increased under inflammatory conditions. The extravasation of leukocytes to the sites of inflammation is comprised of distinct adhesive events. Blockade of these adhesive mechanisms results in a reduced

Immune Mechanisms of Pain and Analgesia, edited by Halina Machelska and Christoph Stein.

migration of opioid-containing immune cells (see ref. 9 and the chapter by Machelska in this volume). Importantly, these potentially antinociceptive (analgesic) opioid peptides can be secreted into the surrounding tissue by specific releasing factors. This results in an activation of opioid receptors on sensory nerve endings and can elicit an inhibition of the generation and transmission of painful stimuli. The best described releasing factors so far are corticotropin-releasing factor (CRF) and IL-1. This book chapter outlines in more detail the role of opioid peptide-releasing factors in endogenous pain control.

Opioid Peptide Release

In the pituitary CRF derived from the hypothalamus activates CRF receptors expressed in corticotrophic cells[10,11] to stimulate the release of proopiomelanocortin (POMC)-derived peptides such as adrenocorticotropic hormone (ACTH), β-endorphin (END) and others.[12] In a similar manner, IL-1 may act on its receptors on corticotrophic cells[13,14] to secrete ACTH and END.[15,16] Other cytokines such as TNF-α as well as IL-6 are known to activate the hypothalamo-pituitary-adrenal axis by similar mechanisms.[17,18]

In in vitro studies CRF and IL-1 are known to induce the secretion of POMC-derived peptides such as END from stimulated immune cells.[19-21] In our experiments we have prepared cell suspensions from popliteal lymph nodes of the inflamed and contralateral noninflamed paws.[8,22,23] The immune cells obtained from the draining lymph nodes of inflamed tissue are stimulated in vivo by a chronic inflammatory process. This pathophysiological in vivo situation resembles the clinical situation much more closely than an in vitro stimulation with agents such as lipopolysaccharide, phytohaemagglutinin, etc.[20] The immune cell suspensions were incubated with increasing concentrations of CRF or IL-1 in the presence or absence of their respective receptor antagonists α-helical CRF and IL-1 receptor antagonist. The END content in the supernatant was measured by a specific radioimmunoassay. The results showed that similar to the known CRF- and IL-1-induced release of END from the pituitary, CRF and IL-1 could also induce the secretion of END from immune cells of popliteal lymph nodes (Fig. 1).[8,23] These cells were mainly lymphocytes but also macrophages. The END release dose-dependently increased with increasing doses of CRF and IL-1. It was dose-dependently anatgonized by the respective antagonists indicating a specific effect mediated by CRF and IL-1 receptors on immune cells.[8,23] There are two described pathways of cellular release of substances, a constitutive and a vesicular. In further experiments we examined the Ca^{2+}-dependence of the END release from immune cells. Our data show that in the absence of extracellular Ca^{2+} the END release was significantly attenuated (Fig. 1).[8] In addition, increasing concentrations of extracellular K^+ could elicit the release of END similar to the CRF- and IL-1-induced release. Ca^{2+}-dependence and K^+ -induction clearly point to a vesicular release of END from immune cells[8] similar to the neurotransmitter release from presynaptic nerve endings.[24] Taken together, CRF and IL-1 act on their respective receptors on immune cells to elicit a vesicular release of opioid peptides such as END from resident immune cells within inflamed tissue. This may result in subsequent activation of opioid receptors on nerve terminals of sensory neurons to inhibit the generation and transmission of painful stimuli.

CRF and IL-1 Receptors on Immune Cells

If CRF and IL-1 induce receptor specific END release from immune cells, there should be CRF and IL-1 receptors on immune cells. It is well known that IL-1 receptors exist on immune cells, particularly on lymphocytes.[2] It is also described that CRF

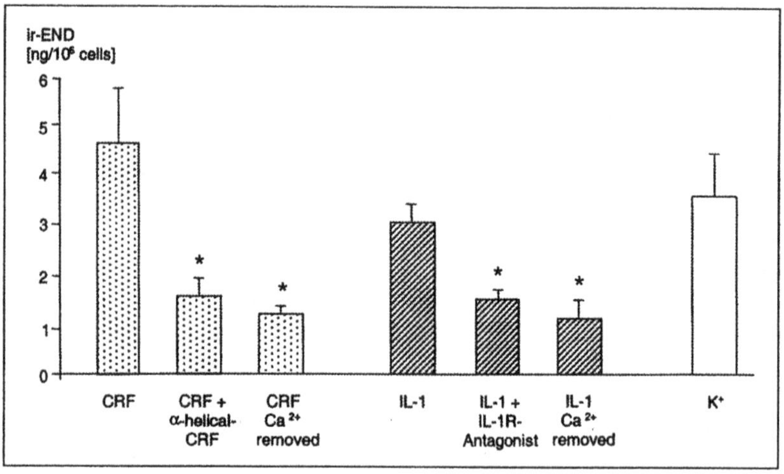

Fig. 1. CRF- and IL-1-induced β-endorphin release in cell suspensions from popliteal lymph nodes of inflamed hind paws. Four days after CFA-hindpaw inflammation popliteal lymph nodes were removed, cell suspensions prepared and incubated with 100 ng CRF (dotted columns) and 100 ng IL-1 (striped columns) with or without their respective antagonists α-helical CRF (100 ng) and IL-1 receptor(IL-1R) antagonist (100 ng). The β-endorphin release into the supernatant was determined by radioimmunoassay. CRF and IL-1 induced a release of β-endorphin into the supernatant. These effects were dose-dependent (see the chapter by Machelska in this volume) and antagonized by the respective antagonists ($P<0.05$, Mann-Whitney U-test). The CRF- and IL-1-induced β-endorphin release was Ca^{2+}-dependent, since removal of Ca^{2+} (0.1 mM EGTA) abolished this release ($P<0.05$, Mann-Whitney U-test). A similar β-endorphin release could be evoked with a 50 mM concentration of a K^+ solution ($P<0.05$, Mann-Whitney U-test). Thus, β-endorphin release from immune cells of popliteal lymph nodes is Ca^{2+}-dependent and K^+-inducible similar to the vesicular release of neurotransmitters. Data represent means ± s.e.m. CRF, corticotropin-releasing factor; IL-1, interleukin-1, CFA, complete Freund's adjuvant.

receptors are abundant within the central nervous system, particularly in the pituitary.[25] In addition to these central CRF receptors, autoradiographic studies could identify CRF receptors in various peripheral tissues[26] and on immune cells such as monocytes/macrophages.[25,27] Only recently following the molecular cloning of the CRF-1 and CRF-2 receptors, CRF receptors can be identified and characterized by specific antibodies.[28]

In the model of complete Freund's adjuvant (CFA)-inflammation of a rat hindpaw, we have examined the presence of CRF- and IL-1 receptors within inflamed versus noninflamed paw and lymph node tissue (ref. 29 and the chapter by Mousa in this volume). Binding sites for the radiolabelled ligands [^{125}J]-CRF and [^{125}J]-IL-1 in lymph nodes and paw tissue were visualized by autoradiography and their density calculated from the autoradiographic film and/or the photo emulsions.[29] In noninflamed lymph nodes [^{125}J]-CRF and [^{125}J]-IL-1 binding sites were only very low abundant. In contrast, in inflamed lymph nodes [^{125}J]-CRF and [^{125}J]-IL-1 binding sites were highly abundant. These binding sites were exclusively located in the medulla of the lymph node tissue which is mainly characteristic for T-lymphocytes. Consistently, in the photoemulsion [^{125}J]-CRF and [^{125}J]-IL-1 binding sites were mainly located at the cell surface of morphologically identified lymphocytes. The autoradiographic examination of noninflamed paw tissue revealed [^{125}J]-CRF and [^{125}J]-IL-1 binding sites neither within subcutaneous paw tissue nor on peripheral nerve fibres.[29] In inflamed paw tissue [^{125}J]-CRF and [^{125}J]-IL-1 binding sites were highly up-regulated.[29] Thus, CRF and IL-1 receptors are not present in normal paw tissue in which inflammatory cells are absent. In contrast, in inflamed paw tissue CRF and IL-1 receptors are highly abundant mainly on the cell surface of inflammatory cells. In line with these findings, the dark field microscopy revealed that [^{125}J]-CRF and [^{125}J]-IL-1 binding sites were primarily at the site of inflammatory foci within the rat hindpaw. In the photoemulsion [^{125}J]-CRF and [^{125}J]-IL-1 binding sites were mainly located at the cell surface of morphologically identified lymphocytes and macrophages. In competition experiments, the pharmacological characterization of the [^{125}J]-CRF and [^{125}J]-IL-1 binding sites showed a B_{max} and K_d[29] similar to those previously described for CRF and IL-1 in brain tissue.[25,30]

Our results are in line with more recent studies in which a polyclonal anti-CRF-1 receptor antibody was used for the identification of CRF receptors in mouse spleenic cells.[31] While in naïve mice only a few CRF-1 receptors could be identified in the spleen, lipopolysaccharide stimulation resulted in a 17-fold increase of CRF-1 receptors, mainly on neutrophils, granulocytes and macrophages. Because the CRF used in our studies has a higher affinity for the CRF-1 than for the CRF-2 receptors,[32] the [^{125}J]-CRF binding sites we characterized are most likely of the CRF-1 type.

CRF- and IL-1-Induced Analgesia

Since CRF and IL-1 induce the release of END from immune cells in vitro this might also occur in vivo. In a series of in vivo experiments we have examined whether the local application of CRF or IL-1 elicits an attenuation of inflammatory pain presumably by a release of END from resident immune cells within inflamed tissue. Four days after CFA-inflammation of a rat hindpaw, local administration of CRF and IL-1 produced a significant elevation of paw pressure thresholds (Fig. 2).[23] These analgesic effects increased dose-dependently. They were not due to a systemic effect since the intravenous administration of the maximal effective doses of CRF or IL-1 did not show any significant changes in paw pressure thresholds. Thus, CRF- and IL-1-induced analgesia is peripherally mediated.

These results are in line with other studies suggesting local analgesic effects of CRF,[26,33] but are in contrast to a report stating that IL-1 induces a state of hyperalgesia (ref. 34 and the chapter by Cunha and Ferreira in this volume). The main differences to the latter are that in our studies IL-1 was not injected into healthy but into highly inflamed paw tissue 4 days after FCA inoculation. In this protracted state of a local inflammation, hyperalgesia is already present and has reached a ceiling effect. However, opioid-containing immune cells are potential target for END release and subsequent activation of opioid receptors on peripheral sensory nerve endings which can result in the attenuation of pain. Co-administration of CRF- and IL-1 receptor antagonists dose-dependently reversed the analgesic effects indicating specific effects mediated through activation of local CRF- and IL-1 receptors (Fig. 2). As mentioned previously, these receptors are mainly located on the cell surface of inflammatory cells that have migrated into the inflamed paw tissue. Since CRF- and IL-1 receptors do not appear to be located on peripheral sensory neurons, we hypothesized that the analgesic effects are most likely mediated by a local release of END from immune cells. In further experiments we could show that immunsuppression either by radiation[35] or by the immunosuppressive drug cyclosporine A[23] resulted in a significant reduction of END-containing cells and of CRF- and IL-1-induced analgesia. This strongly suggests that the analgesic effects of CRF and IL-1 are mediated by a release of opioid peptides from immune cells within inflamed tissue.

In further experiments we examined the involvement of opioid peptides in CRF- and IL-1-induced analgesia by specific immunoneutralization within inflamed paw tissue. Before rats received an intraplantar (directly into the paw) injection of either CRF or IL-1, they

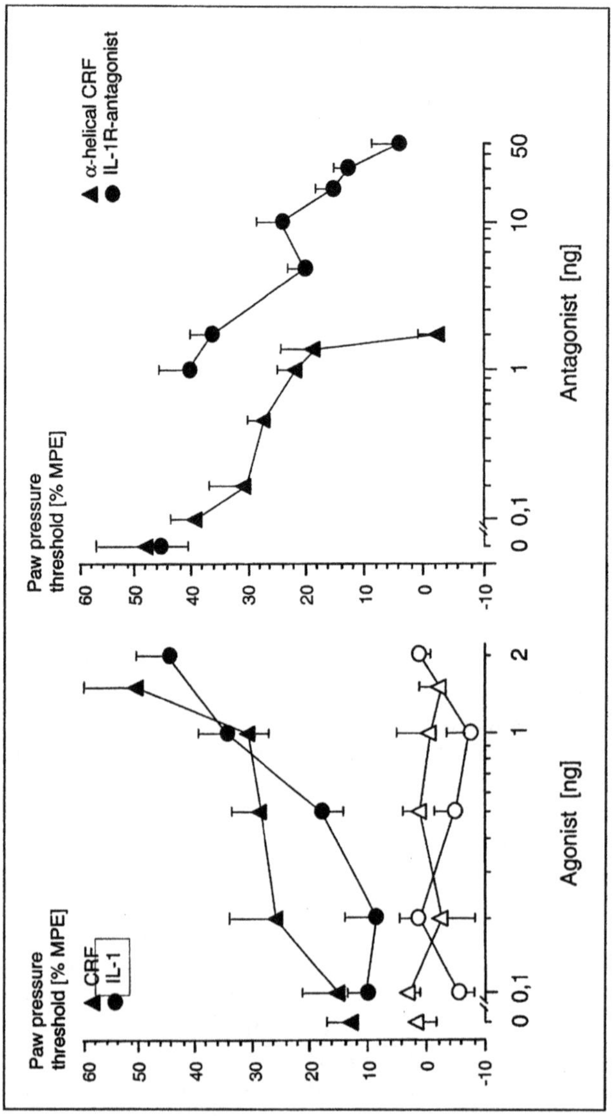

Fig. 2. Analgesic effects of intraplantar CRF (triangles) and IL-1 (circles). Left panel: Intraplantar injections of increasing doses of CRF and IL-1 into CFA inflamed (filled symbols) rat hindpaws elicited significant increases in paw pressure thresholds ($P<0.05$, linear regression ANOVA). These analgesic effects increased dose-dependently ($P<0.05$, linear regression ANOVA). No effects were observed in contralateral noninflamed paws (open symbols). Right panel: CRF-induced analgesia could be dose-dependently reversed by α-helical CRF antagonist (triangles) ($P<0.05$, linear regression ANOVA). IL-1-induced analgesia could be reversed by IL-1 receptor anatgonist (circles) ($P<0.05$, linear regression ANOVA). Data represent means ± s.e.m. CRF, corticotropin-releasing factor; IL-1, interleukin-1, CFA, complete Freund's adjuvant. Reprinted with permission form Schäfer et al. Proc Natl Acad Sci 1994; 91:4219-4223.

were pretreated with a specific antibodies against either END, met-enkephalin or dynorphin. Neutralization of END within inflamed paws resulted in a blockade of CRF- and IL-1-induced analgesia suggesting that END plays a major role (Fig. 3).[23] In addition, met-enkephalin appears to be involved in the CRF- and dynorphin in the IL-1-induced analgesia (Fig. 3).[23]

How do these opioid peptides elicit analgesia? Mu-, δ- and κ-opioid receptors are identified on sensory neurons (see ref. 4 and the chapter by Stein in this volume). We performed in vivo experiments to determine the opioid receptor type involved in CRF- and IL-1-induced analgesia. Our results showed that μ- and δ-opioid receptors appear to be involved in CRF-induced analgesia which is in line with a release of END and

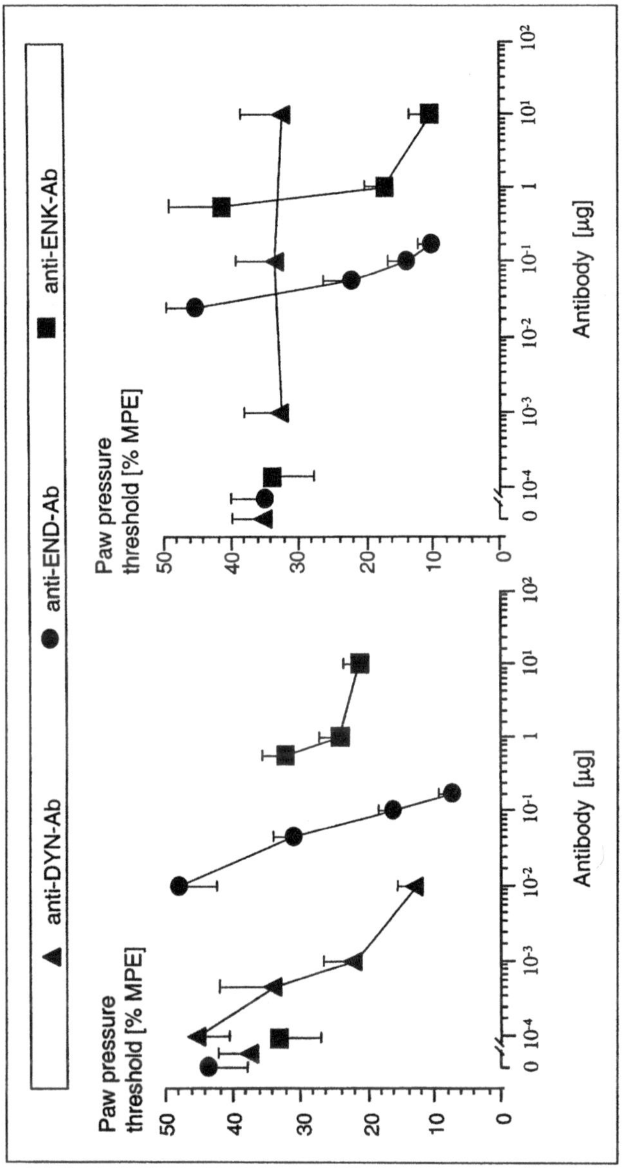

Fig. 3. Blockade of CRF- (left panel) and IL-1- (right panel) induced analgesai by anti-β-endorphin (circles), -met-enkephalin (squares) and -dynorphin (triangles) antibodies. CRF-induced analgesia was dose-dependently reversed by intraplantar anti-β-endorphin and anti-dynorphin antibodies ($P<0.05$, linear regression ANOVA). IL-1-induced analgesia was dose-dependently reversed by intraplantar anti-β-endorphin and anti-met-enkephalin antibodies ($P<0.05$, linear regression ANOVA). Data represent means ± s.e.m. CRF, corticotropin-releasing factor; IL-1, interleukin-1. Reprinted with permission form Schäfer et al. Proc Natl Acad Sci 1994; 91:4219-4223.

met-enkephalin by CRF. Mu-, δ- and κ-opioid receptors seem to be involved in IL-1-induced analgesia according to a release of END and dynorphin by IL-1. Taken together, local injection of certain releasing factors such as CRF and IL-1 trigger the secretion of opioid peptides (mainly END) from resident immune cells within inflamed tissue. Opioid peptides activate their receptors in direct vicinity of the immune cells. Upon activation of these opioid receptors on peripheral nerve terminals of sensory neurons the generation and transmission of painful stimuli is greatly attenuated.

Physiological Relevance of CRF- and IL-1-Induced Analgesia

The question arises whether CRF- and IL-1-induced analgesia is of any physiological relevance. To answer this question we used a specific stress paradigm that was already well characterized in previous experimental studies.[36] In rats with CFA-inflamed hindpaws, a swim stress for 1 min in 4° C cold water elicits a significant increase in paw pressure thresholds in the inflamed paw, however, no changes in the noninflamed paw.[36] This antinociceptive effect is not due to the cooling effect or to central mechanisms since the paw presssure evaluation occurs only in the inflamed paw but not in the contralateral healthy paw. In contrast to the noninflamed paw, opioid peptide-containing immune cells are abundant within inflamed paw tissue. Further experiments showed that opioid peptides such as END produce analgesia because anti-END antibody treatment attenuated this effect.[36] In addition, local injections of naloxone—the traditional opioid receptor antagonist—were able to reverse the stress-induced analgesia[36] (for details see the chapter by Machelska in this volume). This clearly indicates an involvement of peripheral opioid receptors. Thus, 1 min cold water swim stress seems to trigger the release of opioid peptides within inflamed paw tissue which results in activation of peripheral opioid receptors and subsequent potent pain inhibition.

We used this model to further elucidate the endogenous substances that might trigger opioid peptide release. Both CRF and IL-1 are potential candidates for the stress-induced release of END within inflamed tissue. CRF, because it is a well described stress mediator that is released upon stressfull stimuli and leads to a secretion of ACTH but also END from the pituitary.[12] IL-1, because it is highly abundant within inflamed tissue and is also able to induce opioid peptide secretion.[2] In a series of experiments we could show that local α-helical CRF, the CRF-receptor antagonist (Fig. 4), but not IL-1 receptor anatgonist, was able to dose-dependently block stress-induced analgesic effects.[37] Consistently, local anti-CRF (Fig. 4), but not anti-IL-1 antibody treatment dose-dependently attenuated stress-induced analgesia.[37] This suggests that CRF but not IL-1 seems to play an important role in the mechanisms that trigger stress-induced analgesic effects. Most interestingly, local but not circulating CRF seems to play a key role. When we injected CRF up to a hundred-fold higher concentration than circulating CRF plasma levels, we could not mimic the effect of stress-induced analgesia.[37] It is well known that under inflammatory conditions CRF expression is up-regulated within inflamed tissue.[38] We therefore tested the hypothesis, whether locally expressed CRF is involved in stress-induced analgesia. We designed a CRF-specific antisense molecule that was targeted to interrupt local CRF synthesis.[37] Antisense molecules are used to specifically "knock-down" the expression and synthesis of a certain protein. It is hypothesized that antisense molecules interfere with the formation of the translation-initiation start complex and, thus, result in an interruption of protein synthesis (Fig. 5). To verify this, one has to show a significant decrease in the targeted protein level. We could show that CRF-antisense treatment significantly reduced the number of CRF-positive cells within inflamed paw tissue. In addition, we demonstrated that the CRF content within inflamed paws was significantly reduced. Both results confirmed that with our CRF-antisense, but not with a mismatch (control) molecule or saline, CRF expression within inflamed tissue was markedly reduced.[37] Importantly, the local inflammatory process was not influenced. Furthermore, following this CRF-antisense pretreatment, the stress-induced analgesic effect was dramatically reduced (Fig. 4) suggesting that local CRF seems to play a key role in intrinsic endogenous pain control. Such pain control can come from a CRF-induced local release of opioid peptides from immune cells which activate peripheral opioid receptors.

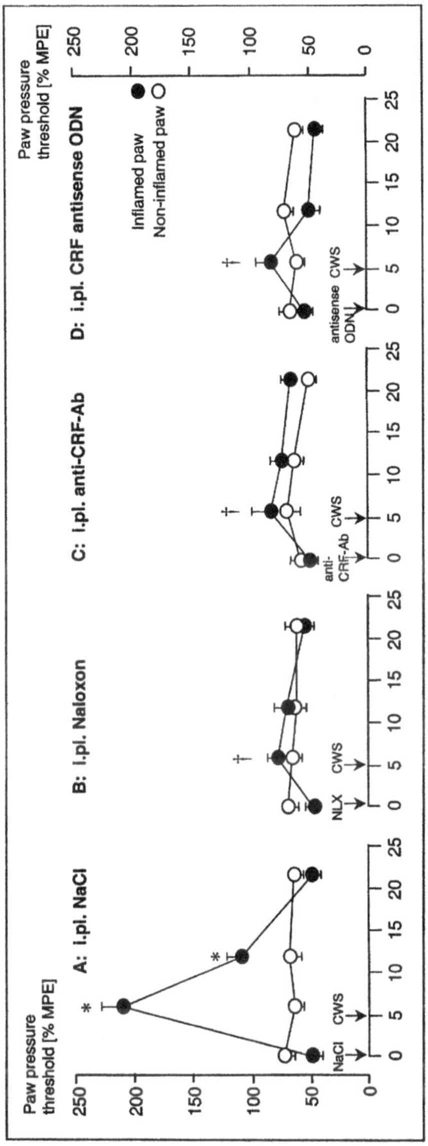

Fig. 4. Swim stress-induced analgesia—involvement of local CRF and peripheral opioid receptors. Cold water (4°C) swim (CWS) stress (for 1 min) induced an elevation of paw pressure thresholds in inflamed (filled circles) but not in noninflamed (open circles) rat paws ($P<0.05$, Friedmann and Wilcoxon test) (A). This analgesic effect was antagonized by intraplantar. naloxone (18 µg) ($P<0.05$, Mann-Whitney U-test), suggesting an involvement of peripheral opioid receptors (B). Neutralization of local CRF by a specific anti-CRF antibody resulted in significant attenuation of the swim stress-induced analgesia ($P<0.05$, Mann-Whitney U-test) (C). Blockade of local CRF synthesis by a specific antisense oligodeoxynucleotide (50 µg/100 µl 30, 18 hours and 6 hours before start of experiments) resulted in the abolishment of swim stress-induced analgesia ($P<0.05$, Mann-Whitney U-test) suggesting that locally expressed CRF plays an important role (D). Data are expressed as means ± s.e.m. CRF, corticotropin-releasing factor; IL-1, interleukin-1.

Taken together, locally expressed CRF, but not cytokines such as IL-1, seem to contribute to physiologically relevant pain control. Such intrinsic pain control mechanisms become apparent in a paradigm of swim stress-induced analgesia. Local treatment with a CRF receptor antagonist, a CRF antibody and a CRF antisense molecule reversed swim stress-induced analgesia indicating that local CRF is an important trigger substance to induce opioid peptide release from resident immune cells within inflamed tissue. This can result in physiologically relevant pain inhibition. Thus, an interaction between the immune and nervous systems appears to be important for both the generation and intrinsic control of pain. Depending on the individual situation one or the other might be more

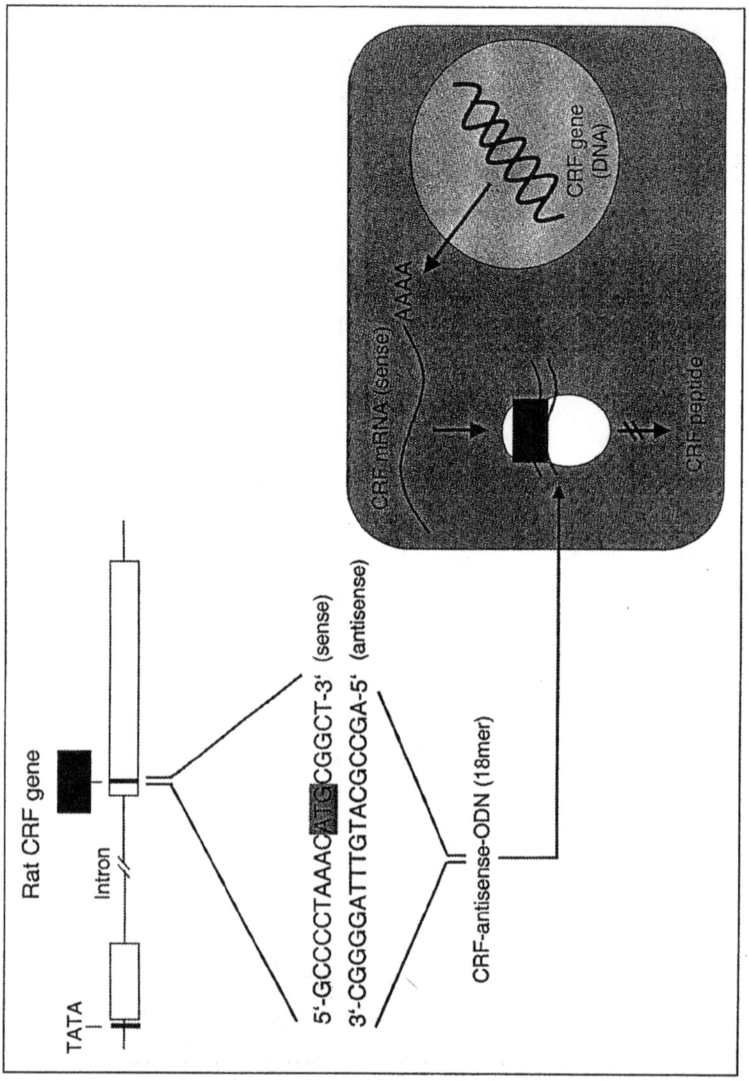

Fig. 5. Possible mechanisms of blockade of corticotropin-releasing factor (CRF) synthesis by specific CRF antisense oligodeoxynucleotides. CRF specific antisense oligodeoxynucleotides were designed corresponding to a nucleotid sequence near the translation-initiation start site of the CRF cDNA. Repeated intraplantar injections of CRF antisense molecules most likely result in the cellular up-take by endocythosis. The intracellular CRF mRNA will be presumably blocked by the incorporated antisense molecules to form a translation-initiation complex. Thus, normal CRF synthesis within the cell is significantly reduced.

prominent. A continuous imbalance in favor of the maintenance of a painful situation might lead to the development of chronic pain.

Summary

Tissue damage causes an inflammatory response in which cytokines contribute to a painful state. Local inflammation also leads to an enhanced expression of opioid peptides such as END within immune cells of inflamed tissue. These endogenous substances can be released by "releasing factors" such as CRF and IL-1 via activation of their receptors on the cell surface of inflammatory cells. Local application of CRF or IL-1 into inflamed tissue results in significant analgesia which is most likely mediated by a release of opioid peptides from immune cells within inflamed tissue. This mechanism of pain inhibition also seems to have a physiological role. Upon certain stressful stimuli analgesic effects seem to be mediated by a release of opioid peptides and a subsequent activation of peripheral opioid receptors. Locally expressed CRF but not IL-1 appear to trigger this release. Thus, inflammatory pain can be modulated both by exogenous CRF and IL-1 as well as endogenous CRF. These mechanisms are based on interactions between the immune and nervous systems. Both the initiation of pain and its control can be regarded as the body's response to prevent further injury, to support wound healing and to return to a normal function as quickly as possible.

References

1. Woolf CJ, Salter MW. Neuronal plasticity: increasing the gain in pain. Science 2000; 288(5472):1765-1769.
2. Dinarello CA, Interleukin-1, interleukin-1 receptors and interleukin-1 receptor antagonist. Int Rev Immunol 1998; 16:457-499.
3. Mansour A, Fox CA, Burke et al. Mu, delta, and kappa opioid receptor mRNA expression in the rat CNS: an in situ hybridization study. J Comp Neurol 1994; 350: 412-38.
4. Schäfer M, Imai Y, Uhl GR et al. Inflammation enhances peripheral mu-opioid receptor-mediated analgesia, but not mu-opioid receptor transcription in dorsal root ganglia. Eur J Pharmacol 1995; 279:165-9.
5. Hassan AH, Ableitner A, Stein C et al. Inflammation of the rat paw enhances axonal transport of opioid receptors in the sciatic nerve and increases their density in the inflamed tissue. Neuroscience 1993; 55:185-95.
6. Mousa SA, Zhang Q, Sitte N et al. beta-Endorphin-containing memory-cells and mu-opioid receptors undergo transport to peripheral inflamed tissue. J Neuroimmunol 2001; 115(1-2):71-78.
7. Stein C, Machelska H, Binder W et al. Peripheral opioid analgesia. Curr Opin Pharmacol 2000, in press.
8. Cabot PJ, Carter L, Gaiddon C et al. Immune cell-derived beta-endorphin. Production, release, and control of inflammatory pain in rats, J Clin Invest 1997; 100:142-148.
9. Machelska H, Cabot PJ, Mousa SA et al. Pain control in inflammation governed by selectins [see comments], Nat Med 1998; 4:1425-8.
10. Wynn PC, Aguilera J, Morell et al. Properties and regulation of high-affinity pituitary receptors for corticotropin-releasing factor. Biochem Biophys Res Commun 1983; 110:602-605.
11. Chen R, Lewis KA, Perrin MH et al. Expression cloning of a human corticotropin-releasing-factor receptor. Proc Natl Acad Sci USA 1993; 90(19):8967-71.
12. Rivier C, Brownstein M, Spiess J et al. In vivo corticotropin-releasing factor-induced secretion of adrenocorticotropin, beta-endorphin and corticosterone. Endocrinology 1982; 110:272-276.
13. Ban E, Milon G, Prudhomme N et al. Receptors for interleukin-1 (alpha and beta) in mouse brain: mapping and neuronal localization in hippocampus. Neuroscience 1991; 43(1):21-30.
14. Cunningham ET Jr, Wada E, Carter DB et al. In situ histochemical localization of type I interleukin-1 receptor messenger RNA in the central nervous system, pituitary, and adrenal gland of the mouse. J Neurosci 1992; 3:1101-14.
15. Bernton EW, Beach JE, Holaday JW et al. Release of multiple hormones by a direct action of interleukin-1 on pituitary cells. Science 1987; 238(4826):519-21.
16. Fagarasan MO, Eskay R, Axelrod J. Interleukin-1 potentiates the secretion of beta-endorphin induced by secretagogues in a mouse pituitary cell line (AtT-20). Proc Natl Acad Sci USA 1989; 86(6):2070-3.
17. Besedovsky HO, del Rey A, Klusman I et al. Cytokines as modulators of the hypothalamus-pituitary-adrenal axis. J Steroid Biochem Mol Biol 1991; 40(4-6):613-8.
18. Perlstein RS, Whitnall MH, Abrams JS et al. Synergistic roles of interleukin-6, interleukin-1, and tumor necrosis factor in the adrenocorticotropin response to bacterial lipopolysaccharide in vivo. Endocrinology 1993;132(3):946-52.
19. Smith EM, Morrill AC, Meyer WJ et al. Corticotropin releasing factor induction of leukocyte-derived immunoreactive ACTH and endorphins, Nature 1986; 321:881-882.
20. Kavelaars A, Ballieux RE and Heijnen CJ. The role of IL-1 in the corticotropin-releasing factor and arginine- vasopressin-induced secretion of immunoreactive beta-endorphin by human peripheral blood mononuclear cells. J Immunol 1989; 142:2338-42.

21. Heijnen CJ, Kavelaars A, Ballieux RE. Beta-endorphin: cytokine and neuropeptide. Immunol Rev 1991; 119:41-63.
22. Cabot PJ, Carter L, Schäfer M et al. Methionine-enkephalin- and Dynorphin A-release from immune cells and control of inflammatory pain. Pain 2001, in press.
23. Schäfer, M, Carter, L, Stein, C. Interleukin-1 beta and corticotropin-releasing factor inhibit pain by releasing opioids from immune cells in inflamed tissue, Proc Natl Acad Sci USA 1994; 91:4219-23.
24. Meir A, Ginsburg S, Butkevich A et al. Ion channels in presynaptic nerve terminals and control of transmitter release. Physiol Rev 1999; 79:1019-88.
25. De Souza EB, Perrin MH, Rivier J et al. Corticotropin-releasing factor receptors in rat pituitary gland: autoradiographic localization. Brain Res 1984; 296(1):202-7.
26. Schäfer M, Mousa SA and Stein C. Corticotropin-releasing factor in antinociception and inflammation, Eur J Pharmacol 1997, 323: 1-10.
27. Webster EL and De Souza EB. Corticotropin-releasing factor receptors in mouse spleen: identification, autoradiographic localization, and regulation by divalent cations and guanine nucleotides. Endocrinology 1988; 122: 609-17.
28. Castro MG, Morrison E. Post-translational processing of proopiomelanocortin in the pituitary and in the brain. Crit Rev Neurobiol 1997; 11(1):35-57.
29. Mousa, S.A., Schäfer, M., Mitchell, WM et al. Local upregulation of corticotropin-releasing hormone and interleukin-1 receptors in rats with painful hindlimb inflammation. Eur J Pharmacol 1996; 311:221-31.
30. Uhl J, Newton RC, Giri JG et al. Identification of IL-1 receptors on human monocytes, J Immunol 1989; 142:1576-81.
31. Radulovic M, Dautzenberg FM, Sydow S et al. Corticotropin-releasing factor receptor 1 in mouse spleen: expression after immune stimulation and identification of receptor-bearing cells. J Immunol 1999; 162:3013-21.
32. McCarthy JR, Heinrichs SC, Grigoriadis DE. Recent advances with the CRF1 receptor: design of small molecule inhibitors, receptor subtypes and clinical indications. Curr Pharm Des 1999; 5(5):289-315.
33. Hargreaves KM, Dubner R, Costello AH. Corticotropin releasing factor (CRF) has a peripheral site of action for antinociception. Eur J Pharmacol 1989, 170:275-9.
34. Ferreira SH, Lorenzetti BB, Bristow AF et al. Interleukin-1 beta as a potent hyperalgesic agent antagonized by a tripeptide analogue. Nature 1988; 334:698-700.
35. Stein C, Hassan AH, Przewlocki R et al. Opioids from immunocytes interact with receptors on sensory nerves to inhibit nociception in inflammation. Proc Natl Acad Sci USA 1990; 87:5935-9.
36. Stein C, Gramsch C and Herz A. Intrinsic mechanisms of antinociception in inflammation: local opioid receptors and beta-endorphin. J Neurosci 1990; 10:1292-8.
37. Schäfer M, Mousa SA, Zhang Q et al. Expression of corticotropin-releasing factor in inflamed tissue is required for intrinsic peripheral opioid analgesia. Proc Natl Acad Sci USA 1996; 93:6096-100.
38. Karalis K, Sano H, Redwine J et al. Autocrine or paracrine inflammatory actions of corticotropin-releasing hormone in vivo. Science 1991; 254:421-3.

CHAPTER 4

Opioid Peptides in Immune Cells

Eric M. Smith

Introduction

The roles of opiates and opioids (endogenous peptides with opiate activity) in the immune system have only recently begun to receive rigorous study. The purpose of this chapter is to cover a poorly understood aspect of the field, the production of opioids by lymphoid cells. Opioid peptides not only affect immune system functions, but mononuclear leukocytes express opioid peptide genes and secrete the active peptides. Lymphoid production of opioids may in part explain many of the conflicting results observed of opioid effects on the immune system. The scope of the discussion will be limited to vertebrate systems, as invertebrates are covered in the chapter by Stefano in this book. The goal is to cover the initial work in the area, particularly our own and to review the latest findings on the topic. Finally, I hope to synthesize the information in terms of its implications and future directions.

Opioids and Opioid Receptors

At the time of the discovery that the immune system produces endogenous opioids,[1] recent work had shown that opiates inhibited immune functions and the initial studies of direct effects of opiates and opioids on lymphocytes were just beginning. The finding of new genes and members of the opioid family of peptides plus characterization of their processing from common precursors stimulated interest as to the effects of these peptides in immune function. Logically, the examination of opioids in the immune system followed the findings with the nervous system. Morphine is the principal alkaloid responsible for the beneficial actions and also the undesirable side effects of opium (for review see ref. 2). Pharmacological studies showed that morphine and related synthetic compounds' effects such as analgesia and addiction liability were stereospecific and mediated by multiple receptors. Four major classes of opioid receptors have been identified, named for the prototype drugs used: delta (δ, deferens), kappa (κ, ketocyclazocine), mu (μ, morphine), and sigma (σ, SKF 10047). The discovery of specific receptors for opiates present in the nervous system led to exploration of their subtypes, specificity and natural roles. Logically it also led to speculation and a search for endogenous opiate receptor ligands. Three separate, but homologous genes encode large precursor proteins containing the opioid peptides: endorphins (END), enkephalins (ENK), and dynorphin. Methionine (met)- or leucine (leu)- ENK is present in the amino terminus of all these opioids. Proopiomelanocortin (POMC) contains β-lipotropin and is differentially cleaved to α-, β-, and γ- END. Proenkephalin A (PEA) is differentially processed into met- and leu-enkephalin, the first

Immune Mechanisms of Pain and Analgesia, edited by Halina Machelska and Christoph Stein. ©2003 Eurekah.com and Kluwer Academic / Plenum Publishers.

opioid peptides to be identified. PEA also contains met-enkephalin-arg6-phe7, met-enkephalin-arg6-gly7-leu8, and peptide E. Prodynorphin (proenkephalin B) is processed to α-neo-endorphin, β-neo-endorphin, dynorphin A-(1-8), dynorphin A(1-17), leumorphin, and dynorphin B.

The opioid peptides are widely distributed, but concentrated in the brain.[2] Major concentrations of β-END are found in the medial basal hypothalamus and the nucleus of the solitary tract within the medulla oblongata, with β-END containing nerve fibers from these two areas projecting to many areas of the brain. ENK are the most widely distributed opioid peptides which are found in many neuronal systems either in local circuits or transported some distance from the site of synthesis. The level of leu-ENK is lower than that of met-ENK and greatest in the globus pallidus and the remainder of the telencephalon, decreasing in the diencephalon, pons mesencephalon, cerebellum and low concentrations in the hippocampus and cortical areas. Dynorphin peptides generally follow the distribution map for the ENK peptides. The highest concentrations are in the posterior pituitary and the hypothalamus.

The pharmacological effects of opioid peptides are similar to those of plant-derived and synthetic opiate alkaloids. The major physiological roles for endogenous opioids seem to be pain perception, stress mechanisms, respiration, temperature control, as well as diuretic and cardiovascular functions. Behaviorally, they may have a role in tolerance development and physical dependence plus patterns such as sexual behavior, feeding and drinking, grooming, and locomotor and operant behaviors.

Peripheral sites of opioid production have also been identified. The placenta, testis, gut, lung, and lymphoid tissues have all been shown to contain or produce POMC-related peptides.[3] The role of peripheral opioids is an active area of research and the theme of this book. Similar relationships and actions for central and peripheral opioids are present in invertebrates (see chapter by Stefano in this book). This evolutionary conservation suggests a fundamental physiologic role for the opioids.

Opioid Effects on the Immune System

Opiate effects on the immune system (immunomodulation) are a complex topic and covered in detail in a chapter by Sacerdote et al in this book. Initially, anecdotal impressions of poor health and susceptibility to infection in addicts would suggest that opiates are immunosuppressive. This has been borne out in experimental animal studies, including opioid receptor knockout mice,[4] but there are also numerous studies showing opioid enhancement of immune function.[5] The initial in vitro studies by Wybran were the first to suggest that there were specific opiate receptors on lymphocytes and that they were functional.[6] In this case morphine, dextromoramide, levomoramide, and met-enkephalin were tested upon normal human T blood lymphocytes by using the active and total rosette tests. Morphine and dextromoramide inhibited the percentage of active T rosettes. This effect was completely reversed in the presence of a specific antagonist, naloxone. Paradoxically, met-enkephalin increased the percentage of active T rosettes and this effect was also specifically inhibited by naloxone. We found that α- and γ-endorphins, met- and leu-ENK, as well as corticotropin inhibited the in vitro antibody response to sheep red blood cells.[7] Beta-END had no effect on the antibody response. Mathews et al[8] found β-END to enhance cytotoxicity, while McCain et al[9] felt the immune modulating activity was through nonopiate receptors. Initially in vivo effects of opiates and opioids were in conflict with the in vitro studies. ENKs[10] and ENDs[11] enhanced natural killer (NK) cell functions. In contrast, Shavit and colleagues showed opioid production in vivo due to intermittent footshock, significantly decreased survival after MAT 13762B tumor cell injection.[12] The decrease in survival could be reversed with opioid receptor antagonists and morphine mimicked the tumor-enhancing effect of intermittent footshock.[12-13] The mechanism was thought to be through inhibition of NK cell activity.[14]

The controversial effects of opiates and opioids on immune function could be due to many reasons. First is the variety of ligands and receptors. It is not clear exactly what receptor/subtype are present on what lymphocytes. Also, there is much evidence to suggest that subtypes of opioid receptors exist and function differently (see in chapter by Stefano in this book). The discovery that the CALLA moiety on granulocytes and certain lymphocytes, the enkephalinase enzyme, and neutral endopeptidase are all one and the same enzyme suggests that enzymatic regulation of opioid availability could also be a factor.[15] In vivo, secondary processes may mediate the effects such as through the induction of corticosteroids[16] or adrenergic pathways.[17-19] An often overlooked aspect of this would be the kinetics of the various responses. Some nearly instantaneous responses such as nitric oxide (NO) induction occur,[20] so responses measured after multiple hour or day intervals may be due to secondary processes. Lastly, another reason for the conflicting actions of opioids may be that the lymphoid cells are also endogenous producers of opioids.

Leukocyte Production of Opioids

POMC Production by Leukocytes

The first opioid found to be produced by cells of the immune system was endorphin-like based on immunoreactivity, physical properties, and activity.[1] The rationale for examining this stemmed from our finding that lymphocytes produced adrenocorticotropin (ACTH) and it seemed likely that END(s), another cleavage product of POMC would also be present. The initial screening of lymphocytes for END-like substances was accomplished by immunofluorescent staining of Newcastle disease virus (NDV)-infected lymphocytes with antibody to γ-END (Fig. 1). The anti-γ-END antiserum was highly specific, being generated against a synthetic peptide. There was no reactivity between the antiserum and ACTH, uninfected cells or NDV-infected epithelial cells. The kinetics of immunoreactive (ir)-END production seemed to follow that of ir-ACTH; first being detected approximately 6 to 8 hours post infection and peaking at 18 to 24 hours post infection. Paralleling the ir-END production was interferon (IFN)-α synthesis, which also peaked at 18-24 hours post infection. Initially, we had found that anti-γ-END or anti-ACTH antiserum could neutralize antiviral activity in an IFN-α preparation.[21] At that time these results suggested to us the existence of a common molecule composed of IFN-α and some or all of POMC, but this later proved to be incorrect. The neutralization results did lead us to test whether or not IFN had END-like bioactivity such as binding to brain opiate receptors. This was initially determined in vitro by measuring the ability of IFN-α to compete with ^{3}H-dihydromorphine in binding to opiate receptors on the particulate fraction of mouse brain lysate.[1;22] IFN was also able to bind opiate receptors in vivo. Intracerebral injection of as little as 500 U of IFN-α induced an endorphin-like analgesia and catatonia in mice.[22] The fact that both physical binding and biological effects were immediately reversible and preventable by the opiate antagonist, naloxone suggests that the effect be mediated through specific opiate receptors. Naloxone has no antiviral activity by itself nor does it affect IFN-α antiviral activity. Purified IFN-α also had opiate receptor binding and functional properties, which suggests that this activity is innate to the molecule. SDS-polyacrylamide gel analysis showed the END-like activity

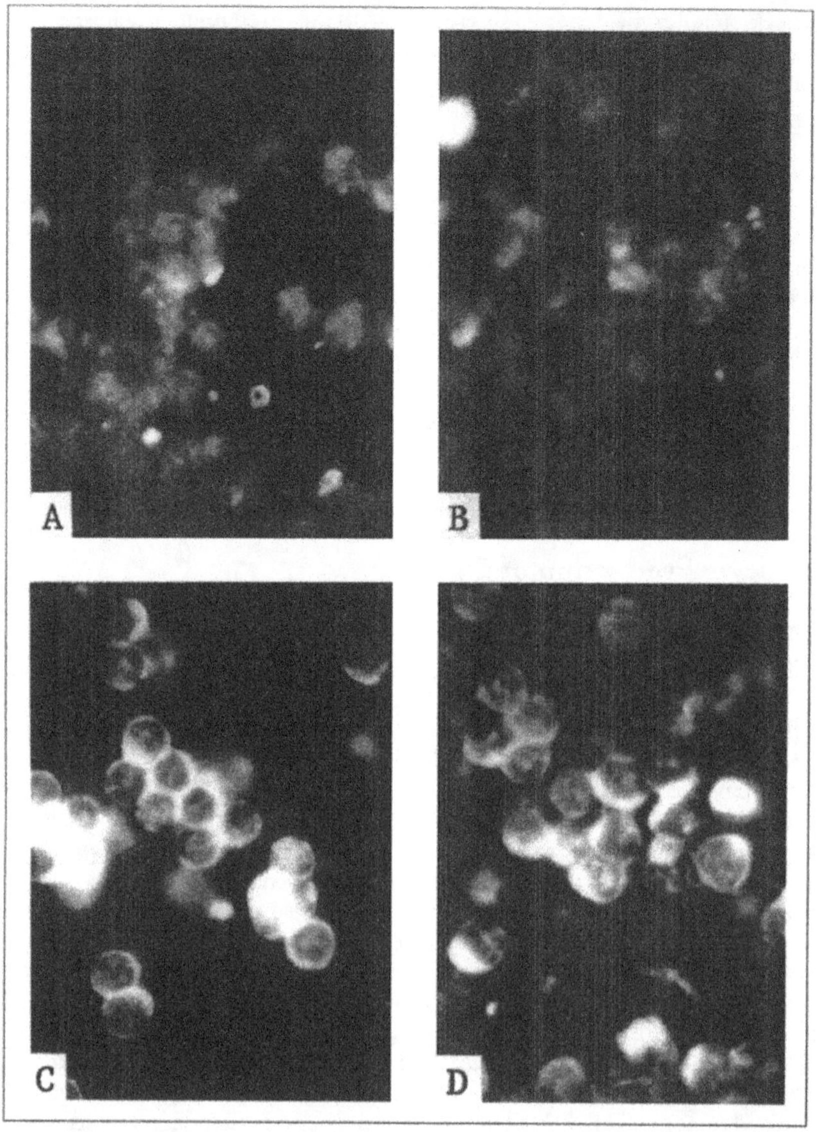

Fig. 1. Immunofluorescence of lymphocyte endorphin production in lymphocytes infected with Newcastle disease virus. (A) Infected lymphocytes at 18 hours after infection stained with normal rabbit serum; (B) noninfected lymphocytes stained with anti-adrenocorticotropin hormone$_{(1-13)}$ antiserum (the fluorescing material in the upper left is nonspecific; (C) infected lymphocytes at 18 hours after infection stained with anti- adrenocorticotropin hormone$_{(1-13)}$ antiserum; and (D) infected lymphocytes at 12 hours after infection stained with anti-γ-endorphin antiserum (X640). Reprinted with permission from Smith EM and Blalock JE. Proc Natl Acad Sci USA 1981; 78(12):7530-7534.

associated principally in two regions of the gel. One region was associated with IFN antiviral activity at a molecular weight range of 18-24 kDa. The majority of the END-like activity was less than 10 kDa and seemed to correspond to β-END (3.6 kDa) or α and γ–END (~1.7 kDa) sized material. Thus it seemed that the END-like activity from NDV-infected lymphocytes can be associated with at least 2 species of molecules, one ir-END and the other IFN-α. Subsequent studies have confirmed the opioid-like activity of IFN-α.[23]

Lolait et al[24] were the first to reproduce our finding of POMC expression by leukocytes at the protein level. Interestingly, they showed immunoreactive β-END in a subpopulation of mouse spleen macrophages with no exogenous stimulation. Ir-ACTH was also produced and as expected nonstimulated, nonmoncytic splenocytes were negative for POMC products. A lot of work characterizing the production and molecular species of POMC products from lymphocytes culminated at the protein level, when we obtained the ACTH amino acid and POMC cDNA sequence from lymphocytes.[25]

POMC mRNA Expression

Westly et al[26] and Lolait et al[27] were the first to show that leukocytes synthesize POMC at the mRNA level. Northern blot analysis revealed that the molecular size of POMC mRNA from NDV-infected murine splenocytes was similar to mRNA from AtT-20 pituitary cells and no POMC specific mRNA could be isolated from noninfected cells.[26] The POMC mRNA detected in the splenic macrophages was also similarly sized (~1100 nt) to pituitary POMC transcripts.[27] This study used HPLC chromatography showed that multiple forms of β-END related peptides were present. These corresponded to α and γ–END and acetylated forms of β-END. Thus, the presence of "immunoreactive" protein could no longer be attributed to adsorption and/or uptake by lymphocytes, there was de novo gene expression and protein synthesis.

The pituitary POMC gene is organized into 3 exons separated by intervening sequences which are removed during processing following transcription to produce the full length, 1200 nt transcript. Expression in nonpituitary tissues has shown production of truncated transcripts as well as full length ones. In particular POMC transcripts found in testes, ovaries, and placenta are all about 200 bases or more shorter than pituitary transcripts.[28-29] Initially, the length of the POMC transcript expressed in leukocytes was reported to be the same size as the pituitary transcript (1200 nt), but a number of investigators subsequently found truncated transcripts. Lacaze-Masmonteil et al[29] found the majority of POMC-related transcripts from the thymus to be approximately 800 nt in length, but there was a small amount of full length mRNA also present. Buzzetti et al[30] found that the transcript was 800 nt from normal peripheral blood mononuclear cells and full length from a transformed T-cell line. In another laboratory, northern blot analysis of total and poly(A)+ RNA demonstrated that human leucocytes contained several POMC mRNA species, including 0.8-, 1.2-, 1.5-, and 9.5-kb transcripts.[31] All four species were expressed in peripheral blood lymphocytes (PBL). The other cell types had either the 0.8-kb species alone or both 0.8- and 9.5-kb species. Neutrophils were the only cells to express the 9.5-kb transcript alone. In T cell clones, interleukin (IL)-2 alone or the antigen for which the clone was specific induced POMC accumulation within 18-24 hours. Cytoplasmic dot blot analysis of PBL RNA demonstrated that POMC expression could be induced by corticotrophin-releasing factor (CRF), recombinant IL-1, and phorbol ester, but not by a calcium ionophore. Galin et al characterized the transcripts induced by CRF in murine lymphocytes.[32-33] By using primers directed at the individual exons, they were able to determine that CRF induced 2 sizes of POMC transcripts. Both were limited to the exon 3, one encoded the entire exon and the second was a smaller exon 3 transcript. This may explain why Mechanick et al[34] detected POMC expression only in monocyte-macrophage lineage cells. These investigators utilized a probe for the POMC exon 1 and this would not have amplified the truncated exon 3-encoding transcripts. However they did detect full-length transcripts in those cells that were positive for β-END. Polymerase chain reaction (PCR)-based cloning, sequencing, and exon mapping of lymphocyte-derived neuroendocrine peptides only showed truncated transcripts that contained the third exon.[35] Why some labs detect primarily truncated transcripts or full length and in what cells is not known, but if probes to exons 1 and 2 were used it may explain those studies that failed to detect any lymphoid POMC expression.[36-37] Lyons and Blalock re-examined the question of POMC mRNA expression using 5' RACE-tailed PCR procedures.[38] Using this exacting and sensitive methodology, they found expression of full length transcripts encoding all three POMC exons. POMC gene expression was detected constitutively in macrophages but only following induction in lymphocytes. Their results unequivocally demonstrated that mononuclear cells can produce full-length POMC transcripts. Its regulation in lymphocytes is distinct from that in macrophages, which constitutively produce POMC-derived peptides and mRNA. Also, the biosynthetic pathway of ACTH from POMC in splenic mononuclear cells stimulated with concanavalin (Con) A appears to be identical to that in rat corticotrophs. Although this does not answer the question about transcript size, it suggests that differences in the inducers, cell type, or other host factors may contribute to the problem.

Inducers of Lymphoid POMC

Leukocyte POMC Modulation by Common Immune System Stimuli

Many stimuli, besides NDV[26;39] can induce POMC in lymphocytes. Infection or transformation with other viruses has been shown to induce POMC expression. Lymphotropic viruses varied in their ability to induce human lymphocytes to produce POMC gene-related transcripts.[40] Epstein-Barr virus (EBV) primarily infects and transforms B-lymphocytes. Examination of cell lines transformed by EBV showed that both adult (K48) and cord blood (Craig) cells had high levels of POMC-transcripts. However, infection of PBL by human T-lymphotrophic viruses (HTLV) I, II, and III (human immunodeficiency virus/HIV) showed no increase in POMC transcription. In contrast, we were able to show that primary infection with HIV I could induce ACTH expression in PBL.[41-42] Possibly our success was due to using a higher multiplicity of infection and longer period of infection.

We found that when all the mononuclear leukocytes (B and T cells, monocytes) in a culture were infected with NDV, they all produced ACTH and END. Activation of specific cell populations by mitogens or other restricted stimuli also induces POMC, but with more restricted expression. Our experience in vivo[43] with endotoxin [bacterial lipopolysaccharide (LPS)] inducing opioid activity in the lymphoid system led us to test LPS in vitro. LPS induced ir-END and ACTH in human PBL.[44] LPS is primarily a B cell and monocyte/macrophage stimulant. We found that in PBL cultures treated with LPS, only 30-40% of the cells, which appear to be small lymphocytes, stained positive by immunofluorescence for ACTH or γ-END.[45] When enriched T or B lymphocyte populations are stained by immunofluorescence, it is clear that LPS primarily induces ir-END production in the B-cell subpopulation (Table 1). The T-cell mitogen, concanavalin A (Con A) stimulates POMC expression from T-cells and biologically active ACTH is secreted from cultures stimulated with either Con A or LPS.[46] Using ACTH production as the indicator, a thorough study of Con A, LPS, and lymphoid cell markers has confirmed

Table 1. Bacterial LPS stimulates γ-endorphin (END) production by B-lymphocytes

		Anti-sera			
Cells	**Treatment**	**NRS***	**γ-END**	**Thy 1.2 (CD90)**	**IgM (μ chain)**
T-Lymphocytes	LPS	4 ± 2#	3 ± 2	N.D.	N.D.
	Mock	6 ± 3	4 ± 2	77 ± 7	N.D.
B-Lymphocytes	LPS	15 ± 8	47 ± 4	N.D.	N.D.
	Mock	4 ± 2	10 ± 4	N.D.	82 ± 10

Enriched cell populations were incubated with lipopolysaccharide (LPS) for 48 hours, harvested and then stained using immunofluorescence with the indicated antisera. *Normal rabbit serum; #positive fluorescing cells ± SD; Mock, medium-treatment; IgM, immunoglobulin M; ND, not determined. Table adapted from reference 45.

that most if not all lymphoid cells are capable of producing POMC peptides given the correct inducer.[47] These results indicate that rat splenic B, cytotoxic T lymphocyte (CTL), and helper T (Th) lymphocytes can be immunologically stimulated to express the peptide hormone ACTH and that basal ACTH expression in macrophages is distinct from that in lymphocytes. There are exceptions, as Staphylococcal enterotoxin A, a T cell mitogen, instead of inducing POMC expression it stimulated thyrotropin production.[48]

Leukocyte POMC Modulation by Neuroendocrine System Products

CRF stimulates pituitary POMC production while corticosteroids are inhibitory. Since lymphocytes have glucocorticoid receptors and corticosteroids are potent anti-inflammatory agents, we determined the effects of a synthetic corticosteroid, dexamethasone on lymphoid POMC expression. Both in vivo and in vitro POMC expression by lymphocytes is inhibited by dexamethasone.[49-50] This was another indicator that the prototypical POMC gene expressed in the pituitary was also being expressed by lymphocytes.

The effect of CRF on pituitary β-END secretion is enhanced by arginine vasopressin (AVP) and this is also true for lymphocyte production of ACTH[50] and β-END.[51-53] Dexamethasone inhibited the ACTH and β-END expression induced by CRF and AVP. Kavelaars et al further found that IL-1 played a central role as a mediator in the β-END induction by CRF and AVP. Neither T-cells or monocytes could be stimulated by CRF/AVP, apparently only the B-cells produce β-END. If an antibody against IL-1 were added to the cultures or monocytes removed, β-END production was inhibited.[51] A subcutaneous administration of CRF induced β-END secretion by lymphocytes in rats.[54] There was a coincident enhanced secretion of IL-1 by macrophages with β-END production by lymphocytes from spleen and mesenteric lymph nodes. Therefore, IL-1 appears to mediate the β-END induction in vivo as well as in vitro. Interestingly, IL-1 induces pituitary cells to release ACTH[55] and similarly amplifies the action of CRF/AVP and forskolin induced β-END secretion in the AtT-20 pituitary tumor cell line.[56]

Beta-adrenergic agonists have been shown to modulate the immune response and there appears to be differences among affected cell types and functions.[57] Kavelaars et al found that in vitro β-adrenergic stimulation of lymphocytes induced the release of β-END.[58] The effect of isoprenaline could be mimicked by using forskolin or dibutyryl cyclic adenosine monophosphate (cAMP) to elevate intracellular concentrations of cAMP which activates protein kinase-A (PKA).

Both the similarity and differences in induction of POMC expression in lymphocytes may be mediated in part by common or different intracellular signaling pathways, respectively. Kavelaars et al found two different signaling pathways for the induction of β-END secretion by human PBL.[59] Activation of protein kinase-C (PKC) with a phorbol ester rapidly induced β-END secretion by T-cells as well as the non-T-cell fraction. Stimulation of PKA by addition of dibutyryl-cAMP also induces the secretion of β-END, but occurred only after 18 hours of culture. In their study the mitogen phytohemagglutinin (PHA), anti-CD3 antibody stimulated PKC and β-END production in the T-cell fraction. LPS, IL-1, and formalinized *Staphylococcal aureus* could not induce β-END secretion. After 18 hours of culture the non-T-cell fraction did secrete β-END and it appeared to be through induction of IL-1. Therefore different stimuli, the activation pathway, cell type, and kinetics are all factors in POMC induction.

Enkephalin Production by Leukocytes

ENK production by the immune system is not as well characterized as the END production, although it was first reported in 1986 by Zurawski et al in activated mouse T helper cells.[60] Both the mRNA and met-ENK protein were detected. Preproenkephalin mRNA was next detected in T-cells, macrophages, and mast cells.[61] Several T-cell lymphomas, a mastocytoma and two macrophage cell lines were found to have high levels of preproenkephalin mRNA. Purified natural macrophages and mast cells also possessed easily detectable levels of this mRNA. Characterization of PEA expression in rat lymphocytes showed that normal B cells were expressing the mRNA and LPS or *Salmonella typhimurium* markedly enhanced the expression within 3 hours.[62] In thymocytes expression was entirely dependent upon stimulation with Con A for over 24 hours of incubation. Examination of PEA during thymocyte maturation after activation by Con A showed expression in CD4 (helper T lymphocyte) positive lymphocytes and not CD8 (CTL) or double positive cells.[63] Since these investigators did not find PEA expression in peripheral T cells they felt its role was in thymocyte maturation. IL-1β induced by the Con A seems to regulate PEA expression.[64] The ENK induced in thymocytes modulated its own expression through delta-2-opioid receptors and seem to function as an inhibitor of thymocyte proliferation.[65] Expression and function of PEA has also been demonstrated in murine fetal thymocytes and also appears to inhibit spontaneous proliferation of these cells.[66]

When rats are exposed to LPS, there is an intense and transient expression of PEA in adrenal glands and lymph nodes.[67] Macrophages located within the lymph nodes and chromaffin cells within the adrenal glands were the sources of PEA and adrenaline enhanced this expression. Further characterization showed that this enhancement was via specific adrenergic receptors.[68] In the adjuvant-induced hind paw inflammation PEA was induced as well as POMC.[69] The IL-1 induced in this model and probable adrenergic activation may be the inducers of PEA in this model. Antisense oligonucleotides against PEA reduced the opioid expression and enhanced the proliferation of Con A stimulated splenocytes.[70] Thus, the ENK produced by lymphocytes may be an important regulator of lymphocyte proliferation.

In vivo Leukocyte Opioid Production and Action

The initial in vivo studies focused on the ACTH product of POMC and adrenal corticosteroid induction, but since ENDs are coproducts this also served to facilitate later studies of opioid production. It was known from clinical studies of ectopic ACTH production, that transformed nonpituitary cells such as small cell carcinomas could produce ACTH and that elevated cortisol levels.[71] As described above, NDV induced cortricosterone in hypophysectomized mice.[49] Although controversial, the finding has been reproduced in other systems[72] and observed in naturally occurring human illnesses.[73] In vivo induction of ir-END accompanies ir-ACTH production. Hypophysectomized mice were inoculated with NDV and followed in a time course for glucocorticoid induction and splenocytes stained by immunofluorescence for ACTH and END. Unexpectedly at the time, the synthetic glucocorticoid, dexamethasone, which inhibits pituitary POMC also, blocked the synthesis of ir-ACTH and ir-END by the spleen cells. There was no reduction in circulating IFN levels. The copresence of the two cleavage products and appropriate negative regulation, was further evidence that the POMC gene was being expressed.

Endotoxic Shock

The first in vivo reports of functional, leukocyte opioid production were in endotoxic (septic) shock. Endotoxic shock occurs during gram-negative bacterial sepsis and is usually experimentally induced by injection of LPS. There is a production of proinflammatory cytokines; IL-1, TNF-α, IL-6 among other factors.[74] Uncontrolled endotoxic/septic shock leads to blood vessel dilation and fever, resulting in hypotension and hyperthermia and eventually death due to congestive heart failure. Investigators studying endotoxic shock found that naloxone alleviated endotoxin-induced hypotension and hyperthermia by apparently blocking the effector molecules.[75] At the same time Traber et al had determined that depletion of thymic lymphocytes also would reduce the cardiopulmonary response to large dosages of endotoxin.[76] Being familiar with our work showing END production by lymphocytes , we collaborated to determine whether the opiate action was related to the lymphocytes and if naloxone would affect the cardiopulmonary response to endotoxin. We were able to measure an increase in opiate-like factors in the plasma of the nonlymphocyte depleted endotoxin treated sheep and cardiopulmonary responses were partially blocked by naloxone.[43] This supported a role for leukocyte endorphins in endotoxic shock.

To test the hypothesis that leukocytes may serve as an extrapituitary source of END, Harbour et al examined the ability of LPS to induce END in vitro.[44] Both human PBL and murine splenocytes were induced by 48 hours treatment with LPS to produce ir-ACTH and ir-END. The leukocyte END was the size of α and γ-END, approximately 1800 daltons and bound to brain opiate receptors. Further study showed that B lymphocyte enriched populations produce immunoreactive END that binds to δ opioid receptors and may play a role in endotoxic shock.[45] Furthermore, there appeared to be a novel processing pathway for POMC induced in lymphocytes by endotoxin, probably through induction of a new prohormone-cleaving enzyme.[77] Instead of the full length of 39 amino acid residues, LPS induced ACTH seemingly contained only 24 residues.[25] Comparing LPS sensitive and resistant strains of mice, we were able to show splenic lymphocyte production of an END during endotoxic shock.[78] The mice were injected with either LPS or B-lymphocyte-derived ir-END. The LPS-sensitive mice (C3HeB/FeJ) presented with a severe hypothermic and pathophysiologic response pattern when treated with LPS or with ir-END. The LPS-resistant mice (C3H/HeJ), which were unresponsive to the LPS, however presented with the typical hypothermic and pathophysiologic responses to the ir-END. Immunofluorescent staining of leukocytes in the LPS-treated mice showed significant ir-END present only in the LPS-sensitive mice at a time point preceding onset of the pathophysiologic response pattern. The exact mechanism for the leukocyte END to induce shock symptoms is not known, but may be in part through induction of IL-1 and TNF-α which are certainly major components of endotoxic shock.[79]

Inflammation

A more recent and very exciting area in which leukocyte-derived opioids appear to function is in inflammation. Other chapters in the book cover the clinical features and implications (see chapters by Schäfer, Mousa and Machelska in this book), but it is important to describe here to emphasize the in vivo induction. In these studies inflammation was in rats injected with complete Freund's adjuvant which induced unilateral hindpaw inflammation. Stein et al showed opioids from immunocytes bound receptors on sensory nerves, resulting in inhibiting nociception (pain).[80] This group has thoroughly characterized the production and action of opioid peptides. Beta-END is detected in the synovial fluid and leukocytes of inflamed joints.[81-83] Synthesis of β-END and met-ENK was demonstrated by B and T lymphocytes at the protein level and expression of POMC and proenkephalin mRNA.[69] Total body irradiation or immunosuppression suppresses the antinociceptive (analgesic) effect in the inflamed paw, which further demonstrated that the opioids were synthesized de novo.

Many of the factors found to induce opioid production by leukocytes in vitro are present in inflamed tissues, particularly cytokines. CRF with or without coproduction of IL-1 is produced at the site and both appear to be inducers.[84-86] Immunohistochemical methods show lymphoid cells and peripheral fibers are producing β-END, met-ENK, dynorphin, and other factors.[86-88] Migration of the opioid-containing immunocytes to the sites of inflammation seems to be mediated in part through co-expression with adhesion molecules.[89] Furthermore, there is intriguing data to suggest that other neurally derived factors may down regulate the opioid response.[90] In a related animal model, adjuvant-induced arthritis, spleen POMC mRNA and IL-β increased.[91] Characterization of this model showed elevated ACTH, β-END, and CRF levels in the spleen and thymus of arthritic animals.[92-93] These neuropeptides seemed to be under negative control of glucocorticoids and in other experiments antisense oligonucleotides which block CRF expression also blocked the splenocyte proliferation.[94] Therefore it appears that factors, such as IL-1 and CRF which are induced in leukocytes and released from peripheral cells induce POMC and PEA production in inflammation.

Human and Other Studies

Descriptions of leukocyte opioid production in human clinical conditions are less common. Paneri and his colleagues have reported β-END production by lymphocytes in several syndromes. Some of these have an inflammatory component such as Alzheimer's disease,[95] multiple sclerosis,[96] rheumatic diseases,[97] Crohn's disease,[98] myocardial infarction,[99] and in experimental allergic encephalomyelitis.[100] In most of these situations, β-END production was increased. In the cases of rheumatic and Crohn's diseases the levels were actually decreased.[97-98] Interestingly, other conditions in which leukocyte β-END production was detected were psychological and stress. These included: eating disorders,[101] depression,[102] schizophrenia,[103] autism,[104] and an experimental inescapable footshock stress in rats.[105-106]

In experimental situations intracerebroventricular injection of IL-1α caused similar elevations of immunocyte β-END production as the foot shock.[105] The IL-1α increases in immunocyte β-END involved CRF, catecholamines, and serotonin. Intermittent but not continuous inescapable footshock stress affected the immune responses and immunocyte β-END concentrations in the rat.[106-107] Paneri and Sacerdote also investigated and compared the properties of microglial cell clones, obtained from embryonic mouse brain primary cultures immortalized with recombinant retroviruses to macrophage clones similarly obtained. Macrophage clones differed from microglial clones in some functions but shared most of the immunological properties. Interestingly, microglial cells were able to produce β-END, and this production was regulated differently in microglial cell clones when compared with macrophages clones. Although LPS treatment induced an increase in β-END concentration in both cell types, only microglial clones and primary microglial cell cultures respond to the neuroendocrine stimulus CRF. In addition, in these cells, β-END release was regulated by a classical neurotransmitter, such as noradrenaline.[108] The authors proposed that this was evidence of communication between neurons and microglial cells.

Other Opioids, Hormones and Cytokines

Table 2 summarizes the multitude of inducers and the major opioids produced by leukocytes. It is important to note that the same type of cells responding to the same inducers, appear able to produce multiple opioids. Another consideration in these studies is that nonopioid factors may also be induced along with the opioids. These could be cytokines or other neuropeptides. Certainly in the case of β-END, a number of studies including ours have shown coproduction of ACTH and β-END .[1,51,86,91,109] Depending upon the cell and the inducing stimuli, there are well over a hundred cytokines, chemokines, leukotrienes, prostaglandins, and neuropeptides that could be produced by leukocytes. These can have direct and/or indirect effects. Cytokines such as IL-1 frequently induce other cytokines and a host of inflammatory factors plus negative regulators such as IL-10 or corticosteroid hormones.[110-111] IL-10 has many systemic effects, particularly at the level of the hypothalamic-pituitary-adrenal axis. In IL-10 deficient mice, corticosteroid levels are elevated and the mice respond more to stress.[111] This may be due to IL-10's action as an inhibitor of cytokine synthesis and we think it may serve as an endogenous negative regulator of IL-1 in the neuroendocrine system as it is in the immune system. A major new subgroup of cytokines, the chemokines are only now being recognized as having major neuroendocrine and immune actions. By definition, the chemokines mediate lymphocyte trafficking to sites of inflammation in particular. Therefore, these may prove to be important regulatory factors in the inflammatory models of opioid production. When we inject gp120 (HIV envelope protein) into rat brains, their sleep is increased[112] plus chemokines and their receptors are expressed at the same time (Hutton, Hogan, Smith, Opp, submitted for publication). Gp120 also induces lymphocyte POMC production,[41] it is highly likely that these processes are related and co-exist.

There are now at least 23 neuropeptide hormones or neurotransmitters found to be produced by cells of the immune system.[110] Many of these are induced by similar stimuli, such as activation with the mitogen Con A. As with the cytokines, lymphoid production of substance P, calcitonin gene-related peptide, vasoactive intestinal peptide, and neuropeptide Y have all been associated with inflammation or hypersensitivity responses and could act in conjunction with leukocyte-derived opioids. Detection of CRF expression in immunocytes has a direct bearing on induction of opioids.[109] When an antisense oligonucleotide complementary to CRF mRNA was utilized it inhibited rat splenocyte proliferation.[94] Thus, it does not take extraordinary, circulating levels of inducing factors such as CRF, it can be present and functional at the local site of an immune response.

While END and ENK are the primary opioids produced by the immune system there is new evidence for others and these may garner more interest in the future. Hassan et al

Table 2. Major opioids produced by leukocytes

Opioid	Cell Type	Stimuli	Citation
Endorphin	B, T, M	Viruses, LPS, Con A, PHA, CRF, IL-1	(1;24;46;50;51)
Enkephalin	B, T, M, Mast cells	LPS, Con A, IL-1, Adrenergic	(60;61;64;67-69)
Dynorphin	?	Inflammation	(88)

B and T lymphocytes; M, monocytes/macrophages; LPS, lipopolysaccharide; Con A, concanavalin A; PHA, phytohemagglutinin; CRF, corticotropin-releasing factor; IL, interleukin.

found that dynorphin was detectable within inflammatory cells and cutaneous nerves at the site of Freund's adjuvant-induced hindpaw inflammation.[88] Jessop et al report the novel opioid peptides endomorphin-1 and –2 are present in spleen and thymus. Another new opioid, prepro-nociceptin/OFQ (N/OFQ), and its receptor, nociceptin/OFQ receptor were detected by reverse transcriptase PCR in porcine thymus, lymph nodes, spleen, and isolated splenocytes.[113-114] Curiously, the receptor protein is not expressed in a functional form so the immunocytes themselves may not be the cellular targets for N/OFQ.[113] In summary, the production of opioids by the immune system is only part of a large arena of regulatory substances. There is much work to be done before we can really hope to understand the major physiologic roles of these regulators.

Implications and Future Directions

The practical and clinical implications of opioid production by the immune system are only beginning to become apparent. In particular, the in vivo findings by Stein and Jessop and their colleagues of leukocyte opioid production and local analgesic effects portend a very exciting future. Intuitively, it makes sense that as part of their role in the inflammatory response the leukocytes could release substances that decrease the pain of the injury or response. Certainly it would make evolutionary sense if it resulted in maintaining the ability to hunt for food or flee enemies. Table 3 summarizes the major opioid receptors and actions on leukocytes and provides a perspective to the global nature of the effects.

The implications of opioid production by leukocytes are both academic and practical. These findings are providing a more complete understanding of the multiple factors and complexity involved in inducing and regulating an immune response. Certainly, it underscores the fact that the immune system is not an autonomous self-regulating entity as once thought. This understanding may have practical clinical implications as well. It is easy to conceive of using appropriate treatment with opioid antagonists or agonists to facilitate healing or to inhibit negative components such as shock.

Future studies should look at other actions that lymphoid opioids may mediate, both on immune cells and nonimmune cells at the sites of action. Since the concentrations are relatively low, action is most likely localized within a microenvironment. We suggested a number of years ago, that parasite production of POMC products (ACTH and β-END) may facilitate the infection by suppressing host resistance[115] and it is intriguing to wonder if the β-END may also reduce any pain and therefore sensation associated with large organisms penetrating tissues. Are there other pathogens that may utilize opioids in their

Table 3. Opiate receptors on leukocytes

Opioid Receptor (Prototype Ligand)	Cell Type	Effected Functions	Molecular Identification
μ (Endorphin)	T, M, NK, PMN	Antibody, NK, Proliferation, Chemotaxis, IFN-γ	(116;117)
δ (Enkephalin)	T, M, NK, PMN	Proliferation, NK, Phagocytosis	(118)
κ (Dynorphin)	T, M	Antibody	(119)
Nociceptin/OFQ (Nociceptin/OFQ)	T	Proliferation, Antibody	(113;114)

T lymphocytes; M, monocytes/macrophages; NK, natural killer cells; PMN, polymorphonuclear leukocytes; IFN, interferon. See text for interpretation.

pathogenesis? Therefore it will be important to continue identifying the actions mediated by opioids and new opioids. In this regard, the cloning and genetic knocking out of opioid receptors should be a useful approach to identifying the essential actions of these molecules. Finally, it will be important to do more human clinical studies to verify the relevance of leukocyte-derived opioids and to determine how best to utilize them.

Acknowledgments

This work was supported in part by the National Institutes of Health Grant NS39974.

References

1. Smith EM, Blalock JE. Human lymphocyte production of corticotropin and endorphin-like substances: association with leukocyte interferon. Proc Natl Acad Sci U S A 1981; 78(12):7530-7534.
2. Simon EJ, Hiller JM. Opioid peptides and opioid receptors. In: Siegel GS, Agranoff BW, Albers RW, Molinoff PB, editors. Basic Neurochemistry: Molecular, Cellular, and Medical Aspects. New York: Raven Press, 1994:321-339.
3. Krieger DT. Brain peptides: what, where, and why? Science 1983; 222(4627):975-985.
4. Gaveriaux-Ruff C, Matthes HW, Peluso J et al. Abolition of morphine-immunosuppression in mice lacking the mu-opioid receptor gene. Proc Natl Acad Sci USA 1998; 95(11):6326-6330.
5. Carr DJ, Rogers TJ, Weber RJ. The relevance of opioids and opioid receptors on immunocompetence and immune homeostasis. Proc Soc Exp Biol Med 1996; 213(3):248-257.
6. Wybran J, Appelboom T, Famaey JP et al. Suggestive evidence for receptors for morphine and methionine-enkephalin on normal human blood T lymphocytes. J Immunol 1979; 123(3):1068-1070.
7. Johnson HM, Smith EM, Torres BA et al. Regulation of the in vitro antibody response by neuroendocrine hormones. Proc Natl Acad Sci USA 1982; 79(13):4171-4174.
8. Mathews PM, Froelich CJ, Sibbitt WL et al. Enhancement of natural cytotoxicity by beta-endorphin. J Immunol 1983; 130(4):1658-1662.
9. McCain HW, Lamster IB, Bozzone JM et al. Beta-endorphin modulates human immune activity via nonopiate receptor mechanisms. Life Sci 1982; 31(15):1619-1624.
10. Faith RE, Liang HJ, Murgo AJ et al. Neuroimmunomodulation with enkephalins: enhancement of human natural killer (NK) cell activity in vitro. Clin Immunol Immunopathol 1984; 31(3):412-418.
11. Kay N, Allen J, Morley JE. Endorphins stimulate normal human peripheral blood lymphocyte natural killer activity. Life Sci 1984; 35(1):53-59.
12. Lewis JW, Shavit Y, Terman GW et al. Stress and morphine affect survival of rats challenged with a mammary ascites tumor (MAT 13762B). Nat Immun Cell Growth Regul 1983; 3(1):43-50.
13. Lewis JW, Shavit Y, Terman GW et al. Apparent involvement of opioid peptides in stress-induced enhancement of tumor growth. Peptides 1983; 4(5):635-638.
14. Shavit Y, Lewis JW, Terman GW et al. Opioid peptides mediate the suppressive effect of stress on natural killer cell cytotoxicity. Science 1984; 223(4632):188-190.

15. Shipp MA, Stefano GB, D'Adamio L et al. Down-regulation of enkephalin-mediated inflammatory response by CD10/neutral endopeptidase-24.11 ("enkephalinase"). Nature 1990; 347:394-396.
16. Freier DO, Fuchs BA. A mechanism of action for morphine-induced immunosuppression: corticosterone mediates morphine-induced suppression of natural killer cell activity. J Pharmacol Exp Ther 1994; 270(3):1127-1133.
17. Carr DJ, Mayo S, Gebhardt BM et al. Central alpha-adrenergic involvement in morphine-mediated suppression of splenic natural killer activity. J Neuroimmunol 1994; 53(1):53-63.
18. Fecho K, Dykstra LA, Lysle DT. Evidence for beta adrenergic receptor involvement in the immunomodulatory effects of morphine. J Pharmacol Exp Ther 1993; 265(3):1079-1087.
19. Hall DM, Suo JL, Weber RJ. Opioid mediated effects on the immune system: sympathetic nervous system involvement. J Neuroimmunol 1998; 83(1-2):29-35.
20. Stefano GB, Salzet B, Rialas CM et al. Morphine- and anadamide-stimulated nitric oxide production inhibits presynaptic dopamine release. Brain Res Dev Brain Res 1997; 763(1):63-68.
21. Blalock JE, Smith EM. Human leukocyte interferon: structural and biological relatedness to adrenocorticotropic hormone and endorphins. Proc Natl Acad Sci USA 1980; 77(10):5972-5974.
22. Blalock JE, Smith EM. Human leukocyte interferon (HuIFN-alpha): potent endorphin-like opioid activity. Biochem Biophys Res Commun 1981; 101(2):472-478.
23. Dafny N. Interferon as an endocoids candidate preventing and attenuating opiate addiction. In: Lal H, LaBella F, Lane J, eds. Endocoids. New York: Alan R. Liss, Inc., 1985:269-276.
24. Lolait SJ, Lim AT, Toh BH et al. Immunoreactive beta-endorphin in a subpopulation of mouse spleen macrophages. J Clin Invest 1984; 73(1):277-280.
25. Smith EM, Galin FS, LeBoeuf RD et al. Nucleotide and amino acid sequence of lymphocyte-derived corticotropin: endotoxin induction of a truncated peptide. Proc Natl Acad Sci USA 1990; 87(3):1057-1060.
26. Westly HJ, Kleiss AJ, Kelley KW et al. Newcastle disease virus-infected splenocytes express the proopiomelanocortin gene. J Exp Med 1986; 163(6):1589-1594.
27. Lolait SJ, Clements JA, Markwick AJ et al. Pro-opiomelanocortin messenger ribonucleic acid and posttranslational processing of beta endorphin in spleen macrophages. J Clin Invest 1986; 77(6):1776-1779.
28. Bardin CW, Chen CL, Morris PL et al. Proopiomelanocortin-derived peptides in testis, ovary, and tissues of reproduction. Recent Prog Horm Res 1987; 43:1-28.
29. Lacaze-Masmonteil T, de Keyzer Y, Luton JP et al. Characterization of proopiomelanocortin transcripts in human nonpituitary tissues. Proc Natl Acad Sci USA 1987; 84(20):7261-7265.
30. Buzzetti R, McLoughlin L, Lavender PM et al. Expression of pro-opiomelanocortin gene and quantification of adrenocorticotropic hormone-like immunoreactivity in human normal peripheral mononuclear cells and lymphoid and myeloid malignancies. J Clin Invest 1989; 83(2):733-737.
31. Stephanou A, Fitzharris P, Knight RA et al. Characteristics and kinetics of proopiomelanocortin mRNA expression by human leucocytes. Brain Behav Immun 1991; 5(4):319-327.
32. Galin FS, LeBoeuf RD, Blalock JE. Corticotropin-releasing factor upregulates expression of two truncated pro-opiomelanocortin transcripts in murine lymphocytes. J Neuroimmunol 1991; 31(1):51-58.
33. Galin FS, LeBoeuf RD, Blalock JE. Possible alternate splicing or initiation of the pro-opiomelanocortin gene in lymphocytes. Int J Neurosci 1990; 51(3-4):171-172.
34. Mechanick JI, Levin N, Roberts JL et al. Proopiomelanocortin gene expression in a distinct population of rat spleen and lung leukocytes. Endocrinology 1992; 131(1):518-525.
35. Maier CC, Blalock JE. PCR-based cloning, sequencing, and exon mapping of lymphocyte- derived neuroendocrine peptides. Immunomethods 1994; 5(1):3-7.
36. Olsen NJ, Nicholson WE, DeBold CR et al. Lymphocyte-derived adrenocorticotropin is insufficient to stimulate adrenal steroidogenesis in hypophysectomized rats. Endocrinology 1992; 130(4):2113-2119.
37. van Woudenberg AD, Metzelaar MJ, van der Kleij AA et al. Analysis of proopiomelanocortin (POMC) messenger ribonucleic acid and POMC-derived peptides in human peripheral blood mononuclear cells: no evidence for a lymphocyte-derived POMC system. Endocrinology 1993; 133(5):1922-1933.
38. Lyons PD, Blalock JE. Pro-opiomelanocortin gene expression and protein processing in rat mononuclear leukocytes. J Neuroimmunol 1997; 78(1-2):47-56.
39. Smith EM. Corticotropin and immunoregulation. In: Scharrer B, Smith EM, Stefano GB, eds. Neuropeptides and immunoregulation. Berlin: Springer-Verlag, 1994:28-45.

40. Oates EL, Allaway GP, Armstrong GR et al. Human lymphocytes produce pro-opiomelanocortin gene-related transcripts. Effects of lymphotropic viruses. J Biol Chem 1988; 263(21):10041-10044.
41. Hashemi FB, Hughes TK, Smith EM. Human immunodeficiency virus induction of corticotropin in lymphoid cells. J Clin Endocrinol Metab 1998; 83(12):4373-4381.
42. Smith EM, Hughes TK, Jr., Hashemi F et al. Immunosuppressive effects of corticotropin and melanotropin and their possible significance in human immunodeficiency virus infection. Proc Natl Acad Sci USA 1992; 89(2):782-786.
43. Traber DL, Thomson PD, Blalock JE et al. Action of an opiate receptor blocker on ovine cardiopulmonary responses to endotoxin. Am J Physiol 1983; 245(2):H189-93.
44. Harbour-McMenamin D, Smith EM, Blalock JE. Bacterial lipopolysaccharide induction of leukocyte-derived corticotropin and endorphins. Infect Immun 1985; 48(3):813-817.
45. Harbour-McMenamin D, Smith EM, Blalock JE. B lymphocyte enriched populations produce immunoreactive endorphin that binds to delta opiate receptors and may play a role in endotoxic shock. ICSU Short Report 1986; 4:378-379.
46. Clarke BL, Gebhardt BM, Blalock JE. Mitogen-stimulated lymphocytes release biologically active corticotropin. Endocrinology 1993; 132(3):983-988.
47. Lyons PD, Blalock JE. The kinetics of ACTH expression in rat leukocyte subpopulations. J Neuroimmunol 1995; 63(2):103-112.
48. Smith EM, Phan M, Kruger TE et al. Human lymphocyte production of immunoreactive thyrotropin. Proc Natl Acad Sci USA 1983; 80(19):6010-6013.
49. Smith EM, Meyer WJ, Blalock JE. Virus-induced corticosterone in hypophysectomized mice: a possible lymphoid adrenal axis. Science 1982; 218(4579):1311-1312.
50. Smith EM, Morrill AC, Meyer III WJ et al. Corticotropin releasing factor induction of leukocyte-derived immunoreactive ACTH and endorphins. Nature 1986; 322:881-882.
51. Kavelaars A, Ballieux RE, Heijnen CJ. The role of IL-1 in the corticotropin-releasing factor and arginine- vasopressin-induced secretion of immunoreactive beta- endorphin by human peripheral blood mononuclear cells. J Immunol 1989; 142(7):2338-2342.
52. Kavelaars A, Ballieux RE, Heijnen CJ. Beta-endorphin secretion by human peripheral blood mononuclear cells: regulation by glucocorticoids. Life Sci 1990; 46(17):1233-1240.
53. Kavelaars A, Ballieux RE, Heijnen CJ. Beta-endorphin production by lymphocytes. Int J Neurosci 1990; 51(3-4):173-175.
54. Kavelaars A, Berkenbosch F, Croiset G et al. Induction of beta-endorphin secretion by lymphocytes after subcutaneous administration of corticotropin-releasing factor. Endocrinology 1990; 126(2):759-764.
55. Woloski BM, Smith EM, Meyer III WJ et al. Corticotropin-releasing activity of monokines. Science 1985; 230(4729):1035-1037.
56. Fagarasan MO, Axelrod J. Interleukin-1 amplifies the action of pituitary secretagogues via protein kinases. Int J Neurosci 1990; 51(3-4):311-313.
57. Sanders VM, Baker RA, Ramer-Quinn DS et al. Differential expression of the beta2-adrenergic receptor by Th1 and Th2 clones: implications for cytokine production and B cell help. J Immunol 1997; 158(9):4200-4210.
58. Kavelaars A, Ballieux RE, Heijnen CJ. In vitro beta-adrenergic stimulation of lymphocytes induces the release of immunoreactive beta-endorphin. Endocrinology 1990; 126(6):3028-3032.
59. Kavelaars A, Ballieux RE, Heijnen CJ. Two different signalling pathways for the induction of immunoreactive beta-endorphin secretion by human peripheral blood mononuclear cells. Endocrinology 1991; 128(2):765-770.
60. Zurawski G, Benedik M, Kamb BJ et al. Activation of mouse T-helper cells induces abundant preproenkephalin mRNA synthesis. Science 1986; 232(4751):772-775.
61. Martin J, Prystowsky MB, Angeletti RH. Preproenkephalin mRNA in T-cells, macrophages, and mast cells. J Neurosci Res 1987; 18(1):82-87.
62. Rosen H, Behar O, Abramsky O, Ovadia H. Regulated expression of proenkephalin A in normal lymphocytes. J Immunol 1989; 143(11):3703-3707.
63. Linner KM, Beyer HS, Sharp BM. Induction of the messenger ribonucleic acid for proenkephalin A in cultured murine CD4-positive thymocytes. Endocrinology 1991; 128(2):717-724.
64. Linner KM, Nicol SE, Sharp BM. IL-1 beta modulates the concanavalin-A-induced expression of proenkephalin A mRNA in murine thymocytes. J Pharmacol Exp Ther 1993; 267(3):1566-1572.

65. Linner KM, Quist HE, Sharp BM. Met-enkephalin-containing peptides encoded by proenkephalin A mRNA expressed in activated murine thymocytes inhibit thymocyte proliferation. J Immunol 1995; 154(10):5049-5060.
66. Linner KM, Quist HE, Sharp BM. Expression and function of proenkephalin A messenger ribonucleic acid in murine fetal thymocytes. Endocrinology 1996; 137(3):857-863.
67. Behar O, Ovadia H, Polakiewicz RD et al. Lipopolysaccharide induces proenkephalin gene expression in rat lymph nodes and adrenal glands. Endocrinology 1994; 134(1):475-481.
68. Ovadia H, Magenheim Y, Behar O et al. Molecular characterization of immune derived proenkephalin mRNA and the involvement of the adrenergic system in its expression in rat lymphoid cells. J Neuroimmunol 1996; 68(1-2):77-83.
69. Przewlocki R, Hassan AH, Lason W et al. Gene expression and localization of opioid peptides in immune cells of inflamed tissue: functional role in antinociception. Neuroscience 1992; 48(2):491-500.
70. Fulford AJ, Harbuz MS, Jessop DS. Antisense inhibition of pro-opiomelanocortin and proenkephalin A messenger RNA translation alters rat immune cell function in vitro. J Neuroimmunol 2000; 106(1-2):6-13.
71. Pfluger KH, Gramse M, Gropp C et al. Ectopic ACTH production with autoantibody formation in a patient with acute myeloblastic leukemia. N Engl J Med 1981; 305(27):1632-1636.
72. Bayle JD, Guellati M, Ibos F et al. Brucella abortus antigen stimulates the pituitary-adrenal axis through the extrapituitary B-lymphoid system. Prog Neuroendocrinimmunol 1991; 4:99-105.
73. DuPont AG, Somers G, Van Steirteghem AC et al. Ectopic adrenocorticotropin production: disappearance after removal of inflammatory tissue. J Clin Endocrinol Metab 1984; 58(4):654-658.
74. Dunn AJ. The role of interleukin-1 and tumor necrosis factor alpha in the neurochemical and neuroendocrine responses to endotoxin. Brain Res Bull 1992; 29(6):807-812.
75. Faden AI, Holaday JW. Naloxone treatment of endotoxin shock: stereospecificity of physiologic and pharmacologic effects in the rat. J Pharmacol Exp Ther 1980; 212(3):441-447.
76. Bohs CT, Fish JC, Miller TH et al. Pulmonary vascular response to endotoxin in normal and lymphocyte depleted sheep. Circ Shock 1979; 6(1):13-21.
77. Harbour DV, Smith EM, Blalock JE. Novel processing pathway for proopiomelanocortin in lymphocytes: endotoxin induction of a new prohormone-cleaving enzyme. J Neurosci Res 1987; 18(1):95-101.
78. Harbour DV, Smith EM, Blalock JE. Splenic lymphocyte production of an endorphin during endotoxic shock. Brain Behav Immun 1987; 1(2):123-133.
79. Harbour DV, Galin FS, Hughes TK et al. Role of leukocyte-derived pro-opiomelanocortin peptides in endotoxic shock. Circ Shock 1991; 35(3):181-191.
80. Machelska H, Stein C. Pain control by immune-derived opioids. Clin Exp Pharmacol Physiol 2000; 27(7):533-536.
81. Stein C, Hassan AH, Przewlocki R et al. Opioids from immunocytes interact with receptors on sensory nerves to inhibit nociception in inflammation. Proc Natl Acad Sci USA 1990; 87(15):5935-5939.
82. Stein C, Gramsch C, Hassan AH et al. Local opioid receptors mediating antinociception in inflammation: endogenous ligands. Prog Clin Biol Res 1990; 328:425-427.
83. Stein C, Hassan AH, Lehrberger K et al. Local analgesic effect of endogenous opioid peptides. Lancet 1993; 342(8867):321-324.
84. Czlonkowski A, Stein C, Herz A. Peripheral mechanisms of opioid antinociception in inflammation: involvement of cytokines. Eur J Pharmacol 1993; 242(3):229-235.
85. Mousa SA, Schafer M, Mitchell WM et al. Local upregulation of corticotropin-releasing hormone and interleukin-1 receptors in rats with painful hindlimb inflammation. Eur J Pharmacol 1996; 311(2-3):221-231.
86. Schafer M, Carter L, Stein C. Interleukin-1 beta and corticotropin-releasing factor inhibit pain by releasing opioids from immune cells in inflamed tissue. Proc Natl Acad Sci USA 1994; 91(10):4219-4223.
87. Donnerer J, Schuligoi R, Stein C et al. Upregulation, release and axonal transport of substance P and calcitonin gene-related peptide in adjuvant inflammation and regulatory function of nerve growth factor. Regul Pept 1993; 46(1-2):150-154.
88. Hassan AH, Pzewlocki R, Herz A et al. Dynorphin, a preferential ligand for kappa-opioid receptors, is present in nerve fibers and immune cells within inflamed tissue of the rat. Neurosci Lett 1992; 140(1):85-88.
89. Mousa SA, Machelska H, Schafer M et al. Co-expression of beta-endorphin with adhesion molecules in a model of inflammatory pain. J Neuroimmunol 2000; 108(1-2):160-170.

90. Schafer M, Zhou L, Stein C. Cholecystokinin inhibits peripheral opioid analgesia in inflamed tissue. Neuroscience 1998; 82(2):603-611.
91. Stephanou A, Sarlis NJ, Knight RA et al. Response of pituitary and spleen pro-opiomelanocortin mRNA, and spleen and thymus interleukin-1 beta mRNA to adjuvant arthritis in the rat. J Neuroimmunol 1992; 37(1-2):59-63.
92. Jessop DS, Lightman SL, Chowdrey HS. Effects of a chronic inflammatory stress on levels of pro-opiomelanocortin-derived peptides in the rat spleen and thymus. J Neuroimmunol 1994; 49(1-2):197-203.
93. Jessop DS, Renshaw D, Lightman SL et al. Changes in ACTH and beta-endorphin immunoreactivity in immune tissues during a chronic inflammatory stress are not correlated with changes in corticotropin-releasing hormone and arginine vasopressin. J Neuroimmunol 1995; 60(1-2):29-35.
94. Jessop DS, Harbuz MS, Snelson CL et al. An antisense oligodeoxynucleotide complementary to corticotropin- releasing hormone mRNA inhibits rat splenocyte proliferation in vitro. J Neuroimmunol 1997; 75(1-2):135-140.
95. Panerai AE, Manfredi B, Sacerdote P. Beta-endorphin concentrations in resting peripheral mononuclear cells and after treatment with PHA or serotoninergic drugs in human aging, Alzheimer's disease, and Down's syndrome. Ann N Y Acad Sci 1992; 663:311-318.
96. Gironi M, Martinelli V, Brambilla E et al. Beta-endorphin concentrations in peripheral blood mononuclear cells of patients with multiple sclerosis: effects of treatment with interferon beta. Arch Neurol 2000; 57(8):1178-1181.
97. Wiedermann CJ, Sacerdote P, Mur E et al. Decreased immunoreactive beta-endorphin in mononuclear leucocytes from patients with rheumatic diseases. Clin Exp Immunol 1992; 87(2):178-182.
98. Wiedermann CJ, Sacerdote P, Propst A et al. Decreased beta-endorphin content in peripheral blood mononuclear leukocytes from patients with Crohn's disease. Brain Behav Immun 1994; 8(3):261-269.
99. Buratti T, Schratzberger P, Dunzendorfer S et al. Decreased levels of beta-endorphin in circulating mononuclear leukocytes from patients with acute myocardial infarction. Cardiology 1998; 90(1):43-47.
100. Panerai AE, Radulovic J, Monastra G et al. Beta-endorphin concentrations in brain areas and peritoneal macrophages in rats susceptible and resistant to experimental allergic encephalomyelitis: a possible relationship between tumor necrosis factor alpha and opioids in the disease. J Neuroimmunol 1994; 51(2):169-176.
101. Brambilla F, Brunetta M, Draisci A et al. T-lymphocyte concentrations of cholecystokinin-8 and beta- endorphin in eating disorders: II. Bulimia nervosa. Psychiatry Research 1995; 59(1-2):51-56.
102. Brambilla F, Maggioni M, Panerai AE et al. Beta-endorphin concentration in peripheral blood mononuclear cells of elderly depressed patients—effects of phosphatidylserine therapy. Neuropsychobiology 1996; 34(1):18-21.
103. Panza G, Monzani E, Sacerdote P et al. Beta-endorphin, vasoactive intestinal peptide and cholecystokinin in peripheral blood mononuclear cells from healthy subjects and from drug-free and haloperidol-treated schizophrenic patients. Acta Psychiatrica Scandinavica 1992; 85(3):207-210.
104. Brambilla F, Guareschi-Cazzullo A, Tacchini C et al. Beta-endorphin and cholecystokinin 8 concentrations in peripheral blood mononuclear cells of autistic children. Neuropsychobiology 1997; 35(1):1-4.
105. Sacerdote P, Bianchi M, Manfredi B et al. Intracerebroventricular interleukin-1 alpha increases immunocyte beta-endorphin concentrations in the rat: involvement of corticotropin-releasing hormone, catecholamines, and serotonin. Endocrinology 1994; 135(4):1346-1352.
106. Sacerdote P, Manfredi B, Bianchi M et al. Intermittent but not continuous inescapable footshock stress affects immune responses and immunocyte beta-endorphin concentrations in the rat. Brain Behav Immun 1994; 8(3):251-260.
107. Panerai AE, Sacerdote P, Bianchi M et al. Intermittent but not continuous inescapable footshock stress and intracerebroventricular interleukin-1 similarly affect immune responses and immunocyte beta-endorphin concentrations in the rat. Int J Clin Pharmacol Res 1997; 17(2-3):115-116.
108. Sacerdote P, Denis-Donini S, Paglia P et al. Cloned microglial cells but not macrophages synthesize beta- endorphin in response to CRH activation. GLIA 1993; 9(4):305-310.
109. Stephanou A, Jessop DS, Knight RA et al. Corticotrophin-releasing factor-like immunoreactivity and mRNA in human leukocytes. Brain Behav Immun 1990; 4(1):67-73.
110. Smith EM. Hormonal activities of cytokines. Chem Immunol 1997; 69:185-202.
111. Smith EM, Cadet P, Stefano GB et al. IL-10 as a mediator in the HPA axis and brain. J Neuroimmunol 1999; 100(1-2):140-148.

112. Opp MR, Rady PL, Hughes TKJ et al. Human immunodeficiency virus envelope glycoprotein 120 alters sleep and induces cytokine mRNA expression in rats. Am J Physiol 1996; 270(5 Pt 2):R963-R970.
113. Halford WP, Gebhardt BM, Carr DJ. Functional role and sequence analysis of a lymphocyte orphan opioid receptor. J Neuroimmunol 1995; 59(1-2):91-101.
114. Pampusch MS, Serie JR, Osinski MA et al. Expression of nociceptin/OFQ receptor and prepro-nociceptin/OFQ in lymphoid tissues. Peptides, 2000; 21(12):1865-1870.
115. Duvaux-Miret O, Stefano GB, Smith EM et al. Immunosuppression in the definitive and intermediate hosts of the human parasite Schistosoma mansoni by release of immunoactive neuropeptides. Proc Natl Acad Sci USA 1992; 89(2):778-781.
116. Chuang TK, Killam KF, Jr., Chuang LF et al. Mu opioid receptor gene expression in immune cells. Biochem Biophys Res Commun 1995; 216(3):922-930.
117. Sedqi M, Roy S, Ramakrishnan S et al. Complementary DNA cloning of a mu-opioid receptor from rat peritoneal macrophages. Biochem Biophys Res Commun 1995; 209(2):563-574.
118. Chuang LF, Chuang TK, Killam KF, Jr. et al. Delta opioid receptor gene expression in lymphocytes. Biochem Biophys Res Commun 1994; 202(3):1291-1299.
119. Chuang LF, Chuang TK, Killam KF, Jr. et al. Expression of kappa opioid receptors in human and monkey lymphocytes. Biochem Biophys Res Commun 1995; 209(3):1003-1010.

CHAPTER 5

Opioid Receptors on Peripheral Sensory Neurons

Christoph Stein

Introduction

The interaction of immune cell-derived opioid peptides with opioid receptors on peripheral terminals of primary afferent (sensory) neurons is one of the most extensively investigated immune mechanisms inhibiting pain. Three families of opioid peptides are well characterized within the central nervous and endocrine systems. The major representatives of each family—β-endorphin, met-enkephalin and dynorphin—can interact with three types of opioid receptors—μ, δ and κ—to generate analgesia. In peripheral inflamed tissue these opioid peptides are produced and released from immune cells and activate opioid receptors on sensory nerve terminals.[1] The production and other characteristics of opioid peptides in immune cells will be covered in chapters by Smith, Stefano et al, and Mousa in this volume. This chapter will give an overview of current information on the anatomy and electrophysiology of opioid receptors localized on primary afferent neurons and on the analgesic effects mediated by these receptors. A closely related topic—the antiinflammatory effects mediated by such receptors—will be discussed in the chapter by Walker in this volume.

Anatomy

In the vertebrates primary afferents are pseudounipolar neurons. Their cell body (soma) is localized in the dorsal root ganglion (or in the trigeminal ganglion for cranial nerves). Their central projections terminate in the dorsal horn of the spinal cord and their peripheral ones in the skin (or internal organs), respectively. Peripheral processes of primary afferent neurons are among the longest axons in the body (about one meter in humans) and are usually classified into myelinated (Aδ) and small diameter unmyelinated axons. The latter are also known as C-fibers and are particularly sensitive to the neurotoxin capsaicin, which is frequently used in experiments to selectively eliminate these neurons. Although both fiber types can transmit nociceptive messages from the periphery to the spinal cord, C-fibers are usually considered the dominant "pain" fibers.

Early studies have identified opioid receptors on cell bodies in the dorsal root ganglion and on central terminals of primary afferent neurons within the dorsal horn of the spinal cord.[1] More recently, we and others have detected such receptors also on peripheral processes of primary afferent neurons in animals[2-7] and in humans.[8] Binding experiments

Immune Mechanisms of Pain and Analgesia, edited by Halina Machelska and Christoph Stein.

indicate that the characteristics of opioid receptors on primary afferent neurons are very similar to those in the brain.[3] In 1992 and 1993 three opioid receptors were cloned: the μ-(MOR), the δ-(DOR) and the κ-opioid receptor (KOR).[9] This has made it possible to demonstrate mRNA for all three receptors in the dorsal root ganglion.[10] With the advent of specific antisera, MOR[4,7,11-13], DOR[12-16] and KOR[12,13] were morphologically identified in the dorsal root ganglion and on small diameter primary afferent neurons. Furthermore, we have shown that the molecular mass of the peripheral MOR is identical with that in the central nervous system by Western blot analysis of the rat sciatic nerve (which consists predominantly of primary afferent neurons).[7] Ultrastructural analysis by electron microscopy has revealed DOR on large dense-core vesicles and on the plasmalemma of small diameter primary afferent neuron terminals within the spinal cord.[16] This led the authors to speculate that the activation of opioid receptors on vesicles may be coupled to the release of neuropeptides (e.g., substance P) from primary afferent nerve terminals in the spinal cord. In the rat skin MOR and DOR were detected on peripheral terminals of unmyelinated axons, associated with the axonal membrane, microtubules and mitochondria.[5] This is consistent with the notion that opioid receptors—similar to neuropeptides and other proteins—are systhesized in the dorsal root ganglion and then carried into the central and peripheral terminals via axonal transport (see below). In line with these findings are our functional studies indicating that capsaicin-sensitive (unmyelinated) primary afferent neurons are indeed primarily responsible for the peripheral antinociceptive (analgesic) effects of morphine and of μ-, δ- and κ-selective opioid agonists.[17,18]

It has been suggested that opioid receptors are also located on sympathetic postganglionic neuron terminals and that they may contribute to opioid analgesia.[19] However, the proposed involvement of sympathetic neurons was questioned by other groups.[20,21] Also, studies attempting the direct demonstration of opioid receptor mRNA in sympathetic ganglia have yielded negative results[22] or extremely small amounts.[23] A thorough morphological investigation clearly demonstrated DOR on unmyelinated primary afferent neurons but not on postganglionic sympathetic neurons in skin, lip and cornea.[6] In immunohistochemical and functional studies we have shown that chemical sympathectomy with 6-hydroxydopamine does not change the expression of opioid receptors in the dorsal root ganglion[13] or the peripheral antinociceptive effects of μ-, δ- and κ-opioid agonists in a model of inflammatory pain.[18] Together, these findings have corroborated the notion that peripheral opioid receptors mediating analgesia are exclusively localized on primary afferent neurons.

Electrophysiology

What are the mechanisms leading to analgesia following the activation of such neuronal opioid receptors? Mu-[24-34], δ-[35,36] and κ-[26,37,38] opioid agonists inhibit calcium currents in cultured dorsal root ganglion (or trigeminal ganglion) neurons. In most of these studies opioids were found to preferentially inhibit the high-voltage activated calcium channels. These effects are mediated by G-proteins (G_i and/or G_o).[29,31,32,37,38] It is known that potassium currents are increased by μ- and δ-opioids in neurons of the central nervous system (e.g., locus coeruleus).[39,40] However, in dorsal root ganglion neurons opioid effects on resting or voltage dependent potassium channels could not be detected so far.[41] Thus, the modulation of calcium currents seems to be the primary mechanism for the inhibitory effects of opioids on primary afferent neurons. In addition, the inhibition of a tetrodotoxin-resistant sodium current by a μ-opioid agonist was described.[42]

Provided that these electrophysiological events are similar throughout the neuron, they may underlie the following observations: Opioids attenuate the excitability of the peripheral nociceptive terminal and the propagation of action potentials.[43-45] Similar to their effects at the soma[46,47] and at central terminals[48], opioids inhibit the (calcium dependent) release of excitatory proinflammatory compounds (e.g., substance P) from peripheral sensory nerve endings.[49-51] In addition, morphine was found to inhibit vasodilatation evoked by the antidromic electrical stimulation of C-fibers.[52] All of these mechanisms can result in pain inhibition. Moreover, they may also account for the antiinflammatory actions of opioids (see chapter by Walker in this volume).

Alterations During Inflammation

A large number of studies have shown that peripheral analgesic effects of exogenous opioids are enhanced under inflammatory conditions.[53,54] One possible underlying mechanism is an upregulation, i.e., an increased number of receptors. Opioid receptors are synthesized in the dorsal root ganglion[10,22,53,55] and their expression can be modulated by inflammation in the vicinity of peripheral primary afferent neuron terminals.[12,13] We observed a moderate upregulation of MOR and downregulation of DOR and KOR in dorsal root ganglion[13] while levels of μ-receptor mRNA were not significantly changed.[53] However, several days after the induction of peripheral inflammation, the axonal transport of opioid receptors in fibers of the sciatic nerve is greatly enhanced.[3,7,56] We have shown that the density of opioid receptors on cutaneous nerve fibers in the inflamed tissue increases and that this increase is abolished by ligating the sciatic nerve.[3] These findings indicate that inflammation enhances the peripherally directed axonal transport of opioid receptors which leads to an upregulation on peripheral nerve terminals.

On the other hand, pre-existent, but possibly inactive neuronal opioid receptors may undergo changes owing to the specific milieu (e.g., low pH) of inflamed tissue, and thus be rendered active. Indeed, low pH increases opioid agonist efficacy in vitro by altering the interaction of opioid receptors with G-proteins in neuronal membranes.[57-59] Furthermore, the ability of opioids to decrease the excitability of primary afferent neurons (via inhibition of adenylyl cyclase and subsequent inhibition of cation currents) is much more pronounced when neuronal cyclic adenosine monophosphate (cAMP) levels are increased, a common scenario in inflammation.[60] Together, these findings suggest that a low pH in inflamed tissue may lead to an enhanced functional efficacy of opioid receptors on primary afferent neurons.

Another important question is how opioid peptides (e.g., those derived from immune cells in the vicinity, see chapters by Machelska, Mousa and Schäfer, this volume) reach their receptors on sensory neurons. Under normal circumstances, tight intercellular contacts at the innermost layer of the perineurium (a sheath encasing peripheral nerve fibers) act as a diffusion barrier for high molecular weight or hydrophilic substances such as peptides.[61] However, inflammatory conditions entail a disruption of the perineurial barrier.[61] We demonstrated that this disruption clearly enhances the passage of opioid peptides to sensory neurons.[62] In addition, we found that the number of primary afferent neuron terminals is increased in inflamed tissue ("sprouting").[63] Together, these observations suggest that the access of opioid peptides to opioid receptors on primary afferent neurons is greatly facilitated—if not unrestricted—in inflamed tissue.

Analgesic Effects

This section will concentrate on analgesic effects following the activation of peripheral opioid receptors on primary afferent neurons by exogenous opioid agonists. Effects of endogenous (immune-derived) opioids are discussed in the chapters by Machelska and Schäfer in this volume.

Many conventional opioid agonists produce potent opioid receptor mediated analgesia when administered in small, systemically inactive doses in rodents, primates and humans.[54,64] These effects have been observed mostly under pathological conditions like neuropathic pain[65], colorectal distension[66], bone damage[67] and inflammation[54,64], but also in noninflamed tissue.[62] Central side effects typically associated with systemically (e.g., intravenously) administered opioids like tolerance, sedation, dependence and respiratory depression can be avoided by this peripheral application of agents. Thus, strategies to restrict the access of opioid agonists to the central nervous system have been developed. Such approaches include the incorporation of highly polar hydrophilic substituents or the inclusion of both hydrophilic and hydrophobic portions in molecules. The aim is to achieve peripheral selectivity and high analgesic potency of these compounds.[64,68,69] Recently, novel peptidic k-ligands were identified by positional scanning of a tetrapeptide combinatorial library screened in opioid receptor binding assays. We demonstrated that these ligands are peripherally selective, as shown by the lack of sedative activity after systemic administration, and that they have potent analgesic effects.[70] Moreover, these peptides exhibited significant anti-inflammatory properties, as measured by paw volume and histological signs.[70] These preclinical studies show that both conventional and novel opioid analgesics are useful and available for peripheral administration.

Peripheral opioid actions are undoubtedly of clinical relevance. We have detected opioid receptors on peripheral nerve terminals as well as opioid peptides in human synovia.[8,71] A sizeable body of clinical literature has recently been reviewed and has demonstrated the analgesic efficacy of locally applied opioids in various clinical settings.[72] The most extensively studied clinical situation is the intraarticular application of opioid agonists for pain control after knee surgery.[54,64,72] A recent important development is the analogous use in chronic arthritis.[73,74] Novel routes of administration include the perineural, intra-abdominal, orbital and topical wound infiltration with opioids.[64,72] The majority of these studies has clearly produced evidence for the clinical usefulness of peripheral opioid analgesics. A major goal for the future is the development of peripherally selective opioids which are suitable for oral or intravenous administration in chronic and acute (e.g., postoperative) pain. Preliminary clinical studies are currently testing the oral and intravenous application of peripheral μ- and κ-agonists.[69]

Tolerance

An important question is whether tolerance (i.e., a loss of analgesic efficacy after repeated or continuous application of agonists) does or does not develop at peripheral opioid receptors. Peripheral tolerance has been observed in animal models without inflammation.[75] However, since the number, affinity and coupling efficacy of opioid receptors appear to be enhanced under inflammatory conditions (see above under "alterations during inflammation") these studies do not permit conclusions regarding tolerance in the pathological situation. In another model peripherally (but not centrally) mediated morphine analgesia was reported to be resistant to tolerance development.[76,77] A lack of tolerance development was also shown after repeated local administration of loperamide (a μ-agonist) in a thermal inflammatory model.[78] In the same study systemically applied morphine produced only partial cross-tolerance with loperamide.[78] In clinical studies we found a lack of cross-tolerance between local morphine- and endogenous opioid-induced analgesia.[8] So far the weight of the evidence tends to argue for a relative lack of tolerance development at peripheral opioid receptors under inflammatory conditions. Clearly more basic and clinical investigations are needed to resolve this important issue and to elucidate the underlying mechanisms.

Conclusions

In summary, this chapter discussed developments in the field of peripheral opioid receptors, including localization, signalling pathways, novel peripherally restricted agonists, clinical applications and tolerance development. Major recent findings are the functionally exclusive localization of opioid receptors to primary afferent (but not sympathetic) neurons, the relative lack of tolerance under inflammatory conditions and tetrapeptides as peripherally restricted compounds. Clinical studies have now moved into the field of chronic arthritic pain, a problem of major importance and prevalence. An important long-term goal remains to develop opioid analgesics which exclusively activate peripheral opioid receptors on sensory neurons. Such compounds should be devoid of centrally mediated side effects (e.g., sedation, respiratory depression, tolerance, dependence, addiction) and they should be suitable for the oral and/or intravenous route for the widespread use in patients with acute and chronic pain.

Acknowledgments

Supported by grants from the Deutsche Forschungsgemeinschaft, the International Anesthesia Research Society and the National Institutes of Health. Current industrial collaborators include Ferring B.V. and EpiCept Corp.

References

1. Stein C. The control of pain in peripheral tissue by opioids. N Engl J Med 1995; 332(25):1685-1690.
2. Stein C, Hassan AHS, Przewlocki R et al. Opioids from immunocytes interact with receptors on sensory nerves to inhibit nociception in inflammation. Proc Natl Acad Sci USA 1990; 87:5935-5939.
3. Hassan AHS, Ableitner A, Stein C et al. Inflammation of the rat paw enhances axonal transport of opioid receptors in the sciatic nerve and increases their density in the inflamed tissue. Neuroscience 1993; 55:185-195.
4. Li JL, Kaneko T, Mizuno N. Effects of peripheral nerve ligation on expression of mu-opioid receptor in sensory ganglion neurons: an immunohistochemical study in dorsal root and nodose ganglion neurons of the rat. Neurosci Lett 1996; 214(2-3):91-94.
5. Coggeshall RE, Zhou S, Carlton SM. Opiate receptors on peripheral sensory axons. Brain Res 1997; 764(1-2):126-132.
6. Wenk HN, Honda CN. Immunohistochemical localization of delta opioid receptors in peripheral tissues. J Comp Neurol 1999; 408(4):567-79.
7. Mousa SA, Zhang Q, Sitte N et al. β-Endorphin containing memory-cells and μ-opioid receptors undergo transport to peripheral inflamed tissue. J Neuroimmunol 2001; 115:71-78.
8. Stein C, Pflüger M, Yassouridis A et al. No tolerance to peripheral morphine analgesia in presence of opioid expression in inflamed synovia. J Clin Invest 1996; 98:793-799.
9. Kieffer BL. Recent advances in molecular recognition and signal transduction of active peptides: Receptors for opioid peptides. Cell Mol Neurobiol 1995; 15(6):615-635.
10. Mansour A, Fox CA, Burke S et al. Mu, delta, and kappa opioid receptor mRNA expression in the rat CNS: an in situ hybridization study. J Comp Neurol 1994; 350(3):412-38.
11. Arvidsson U, Riedl M, Chakrabarti S et al. Distribution and targeting of a mu-opioid receptor (MOR1) in brain and spinal cord. J Neurosci 1995; 15(5):3328-3341.

12. Ji R-R, Zhang Q, Law P-Y et al. Expression of μ-, δ-, and κ-opioid receptor-like immunoreactivities in rat dorsal root ganglia after carrageenan-induced inflammation. J Neurosci 1995; 15(12):8156-8166.
13. Zhang Q, Schäfer M, Elde R et al. Effects of neurotoxins and hindpaw inflammation on opioid receptor immunoreactivities in dorsal root ganglia. Neuroscience 1998; 85(1):281-291.
14. Dado RJ, Law PY, Loh HH et al. Immunofluorescent identification of a delta-opioid receptor on primary afferent nerve terminals. NeuroReport 1993; 5:341-344.
15. Zhang X, Bao L, Xu ZQ et al. Localization of neuropeptide Y Y1 receptors in the rat nervous system with special reference to somatic receptors on small dorsal root ganglion neurons. Proc Natl Acad Sci USA 1994; 91(24):11738-42.
16. Zhang X, Bao L, Arvidsson U et al. Localization and regulation of the delta-opioid receptor in dorsal root ganglia and spinal cord of the rat and monkey: evidence for association with the membrane of large dense-core vesicles. Neuroscience 1998; 82(4):1225-42.
17. Bartho L, Stein C, Herz A. Involvement of capsaicin-sensitive neurones in hyperalgesia and enhanced opioid antinociception in inflammation. Naunyn Schmiedeberg's Arch Pharmacol 1990; 342:666-670.
18. Zhou L, Zhang Q, Stein C et al. Contribution of opioid receptors on primary afferent versus sympathetic neurons to peripheral opioid analgesia. J Pharmacol Exp Ther 1998; 286(2):1000-1006.
19. Taiwo YO, Levine JD. Kappa- and delta-opioids block sympathetically dependent hyperalgesia. J Neurosci 1991; 11:928-932.
20. Koltzenburg M, Kress M, Reeh PW. The nociceptor sensitization by bradykinin does not depend on sympathetic neurons. Neuroscience 1992; 46:465-473.
21. Schuligoi R, Donnerer J, Amann R. Bradykinin-induced sensitization of afferent neurons in the rat paw. Neuroscience 1994; 59(1):211-5.
22. Schäfer MKH, Bette M, Romeo H et al. Localization of kappa-opioid receptor mRNA in neuronal subpopulations of rat sensory ganglia and spinal cord. Neurosci Lett 1994; 167:137-140.
23. Buzas B, Cox BM. Quantitative analysis of mu and delta opioid receptor gene expression in rat brain and peripheral ganglia using competitive polymerase chain reaction. Neuroscience 1997; 76(2):479-489.
24. Samoriski GM, Gross RA. Functional compartmentalization of opioid desensitization in primary sensory neurons. J Pharmacol Exp Ther 2000; 294(2):500-9.
25. Xie W, Samoriski GM, McLaughlin JP et al. Genetic alteration of phospholipase C beta3 expression modulates behavioral and cellular responses to mu opioids. Proc Natl Acad Sci USA 1999; 96(18):10385-90.
26. Abdulla FA, Smith PA. Axotomy reduces the effect of analgesic opioids yet increases the effect of nociceptin on dorsal root ganglion neurons. J Neurosci 1998; 18(23):9685-94.
27. Khasar SG, Gold MS, Dastmalchi S et al. Selective attenuation of mu-opioid receptor-mediated effects in rat sensory neurons by intrathecal administration of antisense oligodeoxynucleotides. Neurosci Lett 1996; 218(1):17-20.
28. Rusin KI, Moises HC. μ-Opioid receptor activation reduces multiple components of high-threshold calcium current in rat sensory neurons. J Neurosci 1995; 15(6):4315-4327.
29. Wilding TJ, Womack MD, McCleskey EW. Fast, local signal transduction between the μ opioid receptor and Ca^{2+} channels. J Neurosci 1995; 15(5):4124-4132.
30. Taddese A, Nah S-Y, McCleskey EW. Selective opioid inhibition of small nociceptive neurons. Science 1995; 270:1366-1369.
31. Nomura K, Reuveny E, Narahashi T. Opioid inhibition and desensitization of calcium channel currents in rat dorsal root ganglion neurons. J Pharmacol Exp Ther 1994; 270(2):466-474.
32. Moises HC, Rusin KI, Macdonald RL. μ-Opioid receptor-mediated reduction of neuronal calcium current occurs via a G_o-type GTP-binding protein. J Neurosci 1994; 14(6):3842-3851.
33. Schroeder JE, McCleskey EW. Inhibition of Ca2+ currents by a mu-opioid in a defined subset of rat sensory neurons. J Neurosci 1993; 13:867-873.
34. Schroeder JE, Fischbach PS, Zheng D et al. Activation of mu-opioid receptors inhibits transient high- and low-threshold Ca2+ currents, but spares a sustained current. Neuron 1991; 6:13-20.
35. Acosta CG, Lopez HS. delta opioid receptor modulation of several voltage-dependent Ca(2+) currents in rat sensory neurons. J Neurosci 1999; 19(19):8337-48.
36. Werz MA, Macdonald RL. Heterogeneous sensitivity of cultured dorsal root ganglion neurones to opioid peptides selective for mu- and delta-opiate receptors. Nature 1982; 299:730-733.
37. Wiley JW, Moises HC, Gross RA et al. Dynorphin A-mediated reduction in multiple calcium currents involves a G(o) alpha-subtype G protein in rat primary afferent neurons. J Neurophysiol 1997; 77(3):1338-48.

38. Su X, Wachtel RE, Gebhart GF. Inhibition of calcium currents in rat colon sensory neurons by K- but not mu- or delta-opioids. J Neurophysiol 1998; 80(6):3112-9.
39. North RA, Williams JT, Surprenant A et al. Mu and delta receptors belong to a family of receptors that are coupled to potassium channels. Proc Natl Acad Sci USA 1987; 84(15):5487-91.
40. Miyake M, Christie MJ, North RA. Single potassium channels opened by opioids in rat locus ceruleus neurons. Proc Natl Acad Sci USA 1989; 86(9):3419-22.
41. Akins PT, McCleskey EW. Characterization of potassium currents in adult rat sensory neurons and modulation by opioids and cyclic AMP. Neuroscience 1993; 56:759-769.
42. Gold MS, Levine JD. DAMGO inhibits prostaglandin E_2-induced potentiation of a TTX-resistant Na^+ current in rat sensory neurons in vitro. Neurosci Lett 1996; 212:83-86.
43. Russell NJW, Schaible HG, Schmidt RF. Opiates inhibit the discharges of fine afferent units from inflamed knee joint of the cat. Neurosci Lett 1987; 76:107-112.
44. Andreev N, Urban L, Dray A. Opioids suppress spontaneous activity of polymodal nociceptors in rat paw skin induced by ultraviolet irradiation. Neuroscience 1994; 58:793-798.
45. Su X, Julia V, Gebhart GF. Effects of intracolonic opioid receptor agonists on polymodal pelvic nerve afferent fibers in the rat. J Neurophysiol 2000; 83(2):963-70.
46. Jessell TM, Iversen LL. Opiate analgesics inhibit substance P release from rat trigeminal nucleus. Nature 1977; 268:549-551.
47. Mudge AW, Leeman SE, Fischbach GD. Enkephalin inhibits release of substance P from sensory neurons in culture and decreases action potential duration. Proc Natl Acad Sci USA 1979; 76:526-530.
48. Yaksh TL, Jessell TM, Gamse R et al. Intrathecal morphine inhibits substance P release from mammalian spinal cord in vivo. Nature 1980; 286:155-156.
49. Yaksh TL. Substance P release from knee joint afferent terminals: modulation by opioids. Brain Res 1988; 458:319-324.
50. Yonehara N, Imai Y, Chen J-Q et al. Influence of opioids on substance P release evoked by antidromic stimulation of primary afferent fibers in the hind instep of rats. Regul Pept 1992; 38:13-22.
51. Yonehara N, Takiuchi S. Involvement of calcium-activated potassium channels in the inhibitory prejunctional effect of morphine on peripheral sensory nerves. Regul Pept 1997; 68(3):147-53.
52. Shakhanbeh J, Lynn B. Morphine inhibits antidromic vasodilatation without affecting the excitability of C-polymodal nociceptors in the skin of the rat. Brain Res 1993; 607:314-318.
53. Schäfer M, Imai Y, Uhl GR et al. Inflammation enhances peripheral μ-opioid receptor-mediated analgesia, but not μ-opioid receptor transcription in dorsal root ganglia. Eur J Pharmacol 1995; 279:165-169.
54. Stein C, Schäfer M, Cabot PJ et al. Peripheral opioid analgesia. Pain Reviews 1997; 4:171-185.
55. Maekawa K, Minami M, Yabuuchi K et al. In situ hybridization study of mu- and kappa-opioid receptor mRNAs in the rat spinal cord and dorsal root ganglia. Neurosci Lett 1994; 168:97-100.
56. Jeanjean AP, Moussaoui SM, Maloteaux J-M et al. Interleukin-1β induces long-term increase of axonally transported opiate receptors and substance P. Neuroscience 1995; 68(1):151-157.
57. Childers SR. Opiate-inhibited adenylate cyclase in rat brain membranes depleted of G_s-stimulated adenylate cyclase. J Neurochem 1988; 50:543-553.
58. Rasenick MM, Childers SR. Modification of G_s-stimulated adenylate cyclase in brain membranes by low pH pretreatment: correlation with altered guanine nucleotide exchange. J Neurochem 1989; 53:219-225.
59. Selley DE, Breivogel CS, Childers SR. Modification of G protein-coupled functions by low pH pretreatment of membranes from NG108-15 cells: increase in opioid agonist efficacy by decreased inactivation of G proteins. Mol Pharmacol 1993; 44:731-741.
60. Ingram SL, Williams JT. Opioid inhibition of Ih via adenylyl cyclase. Neuron 1994; 13:179-186.
61. Olsson Y. Microenvironment of the peripheral nervous system under normal and pathological conditions. Crit Rev Neurobiol 1990; 5(3):265-311.
62. Antonijevic I, Mousa SA, Schäfer M et al. Perineurial defect and peripheral opioid analgesia in inflammation. J Neurosci 1995; 15(1):165-172.
63. Hassan AHS, Przewlocki R, Herz A et al. Dynorphin, a preferential ligand for kappa-opioid receptors, is present in nerve fibers and immune cells within inflamed tissue of the rat. Neurosci Lett 1992; 140:85-88.
64. Stein C, Machelska H, Binder W et al. Peripheral opioid analgesia. Curr Opin Pharmacol 2001; 1(1):62-65.

65. Kayser V, Lee SH, Guilbaud G. Evidence for a peripheral component in the enhanced antinociceptive effect of a low dose of systemic morphine in rats with peripheral mononeuropathy. Neuroscience 1995; 64(2):537-45.
66. Gebhart GF, Su X, Joshi S et al. Peripheral opioid modulation of visceral pain. Ann NY Acad Sci 2000; 909:41-50.
67. Houghton AK, Valdez JG, Westlund KN. Peripheral morphine administration blocks the development of hyperalgesia and allodynia after bone damage in the rat. Anesthesiology 1998; 89(1):190-201.
68. Machelska H, Binder W, Stein C. Opioid receptors in the periphery. In: Kalso E, McQuay H, Wiesenfeld-Hallin Z, eds. Opioid Sensitivity of Chronic Noncancer Pain. Seattle: IASP Press, 1999:45-58.
69. Brower V. New paths to pain relief. Nat Biotechnol 2000; 18(4):387-91.
70. Binder W, Machelska H, Mousa S et al. Analgesic and anti-inflammatory effects of two novel kappa opioid peptides. Anesthesiology 2001; 94:1034-1044.
71. Stein C, Hassan AHS, Lehrberger K et al. Local analgesic effect of endogenous opioid peptides. Lancet 1993; 342:321-324.
72. Schäfer M. Peripheral opioid analgesia: from experimental to clinical studies. Curr Opin Anaesth 1999; 12:603-607.
73. Likar R, Schäfer M, Paulak F et al. Intraarticular morphine analgesia in chronic pain patients with osteoarthritis. Anesth Analg 1997; 84(6):1313-7.
74. Stein A, Yassouridis A, Szopko C et al. Intraarticular morphine versus dexamethasone in chronic arthritis. Pain 1999; 83(3):525-532.
75. Kolesnikov Y, Pasternak GW. Topical opioids in mice: analgesia and reversal of tolerance by a topical N-methyl-D-aspartate antagonist. J Pharmacol Exp Ther 1999; 290(1):247-52.
76. Tokuyama S, Inoue M, Fuchigami T et al. Lack of tolerance in peripheral opioid analgesia in mice. Life Sci 1998; 62(17-18):1677-81.
77. Ueda H, Inoue M. Peripheral morphine analgesia resistant to tolerance in chronic morphine-treated mice. Neurosci Lett 1999; 266(2):105-8.
78. Nozaki-Taguchi N, Yaksh TL. Characterization of the antihyperalgesic action of a novel peripheral mu-opioid receptor agonist—loperamide. Anesthesiology 1999; 90(1):225-34.

Chapter 6

Morphological Correlates of Immune-Mediated Peripheral Opioid Analgesia

Shaaban A. Mousa

Introduction

Recent research has shown that effective inhibition of pain by endogenous mechanisms can be generated within peripheral tissue, outside the central nervous system. Studies using sensitive and specific techniques of immunohistochemistry have demonstrated that opioid receptors are present on small diameter neurons in the dorsal root ganglia (DRG). Under inflammatory conditions the expression of opioid receptors in the DRG is enhanced. In addition, these receptors are axonally transported in fibers of the sciatic nerve towards the peripheral sensory nerve terminals. Consequently, the number of opioid receptors on nerve fibers in the inflamed subcutaneous tissue increases and this increase is abolished by ligating the sciatic nerve (see also chapter 5). In parallel, endogenous ligands at opioid receptors (mainly β-endorphin) are synthesized in circulating immune cells which migrate to the injured tissue. The extravasation of immunocytes to sites of inflammation is mediated by adhesion molecules, glycoproteins located on the surface of leukocytes and endothelial cells of the vessels. These mechanisms are also involved in the opioid control of inflammatory pain (see chapter 7).[1] This is supported by findings showing an upregulation of adhesion molecules and their colocalization with β-endorphin in immune cells which have migrated to inflamed tissue. Environmental stressful stimuli and releasing agents (corticotropin-releasing factor, CRF; cytokines) can activate immune cells to release opioid peptides. These bind to opioid receptors localized on peripheral sensory nerve terminals. As a consequence of this interaction, inflammatory pain is inhibited. [2,3] This chapter focuses on the anatomical substrates of the interaction between the peripheral nervous system and opioid-containing immunocytes.

Expression of Opioid Receptors on Peripheral Sensory Neurons

Peripheral antinociceptive (analgesic) effects of exogenous opioids are enhanced under inflammatory conditions (see refs. 2, 4 and chapter 5). One possible underlying mechanism for the increased efficacy of opioid agonists is an up-regulation of opioid receptors. Using a specific antibody against μ opioid receptors (MOR) we were able to immunolocalize MOR on small-diameter neurons in the DRG. Furthermore, we found a significant increase

Immune Mechanisms of Pain and Analgesia, edited by Halina Machelska and Christoph Stein.

in the percentage of MOR positive neurons in the ipsilateral DRG 4 days after induction of hindpaw inflammation with complete Freund's adjuvant (CFA) (Fig. 1).[5,6] This is consistent with a previous report showing a similar increase 3 days after carrageenan-produced peripheral inflammation.[7] The MOR immunoreactive neurons in DRG are small and they are vulnerable to capsaicin, a specific neurotoxin that destroys nociceptive C-fibers and abolishes μ, δ and κ receptor agonist-induced peripheral analgesia.[5,8]

Previous studies have shown that opioid receptors are synthesized in the cell body of sensory neurons in DRG and undergo axonal transport to reach the nerve terminals.[9] Further experiments using autoradiography have demonstrated that inflammation induces an increase in the transport of opioid receptors along the sciatic nerve 4 days after CFA.[10] We extended these studies by use of a specific antibody against cloned MOR. Our present results confirmed an increased, especially anterograde transport of MOR along the sciatic nerve (Fig. 1).[6] At the peripheral sensory nerve terminals we detected MOR-immunoreactivity (IR) in inflamed (Fig. 1) and noninflamed subcutaneous tissue.[6] In addition, these studies clearly showed that MOR positive nerve terminals were more abundant in inflamed subcutaneous tissue.[6] These new findings provide the first direct evidence to support the notion that the number of MOR is increased in inflammatory pain.

The fact that peripheral analgesic effects of opioids are pronounced in inflamed tissue but negligible under normal conditions,[4] suggests an inflammation-induced process. In addition to the enhanced axonal receptor transport, there is evidence that the accessibility of opioid receptors on sensory nerves is facilitated in injured tissue. The detection of MOR-IR at the peripheral nerve terminals in noninflamed tissue is in accord with our previous studies using biotinylated anti-id-14 (a monoclonal antiidiotypic antibody against μ- and δ-opioid receptors), [11] and with functional studies showing fentanyl (μ-agonist)-induced peripheral analgesia in noninflamed tissue.[12] This is supported by other studies showing μ-opioid receptors on unmyelinated cutaneous sensory axons in noninjured tissue.[13] In addition, opioid receptors are located not only at the tips of afferent nerve terminals but also more proximally along the axon.[10] These loci are ensheathed by perineurium [14] and are potential sites of opioid action. Inflammatory conditions entail a deficiency of the perineurial barrier and/or an enhanced permeability of endoneurial capillaries.[14,15] A similar leakage can be produced experimentally by the extraneural application of hyperosmolar solutions.[16] Consistently, our histochemical experiments demonstrated that horseradish peroxidase, applied extraneurally in vivo, does not penetrate into the endoneurium of cutaneous nerves in noninflamed paws but does so at both early and later stages of the inflammatory reaction (Fig. 2).[12] In normal tissue, the perineural administration of either hypertonic saline or mannitol strikingly enhances the passage of horseradish peroxidase into the endoneurium. These anatomical findings strongly support the contention that either inflammatory or artificial disruption of the blood-nerve barrier facilitates the access of macromolecules (e.g., opioid peptides) to sensory neurons.

Expression of Opioid Peptides in Immune Cells

Three families of opioid peptides, ligands at opioid receptors, are well characterized. Each family derives from a distinct gene and precursor protein, proopiomelanocortin (POMC), proenkephalin (PENK) and prodynorphin. Their respective major representative opioid peptides are β-endorphin, met-enkephalin and dynorphin. Each peptide exhibits different affinities and selectivities for the three opioid receptor types μ (β-endorphin, enkephalin), δ (enkephalin, β-endorphin) and κ (dynorphin).[17] Two additional

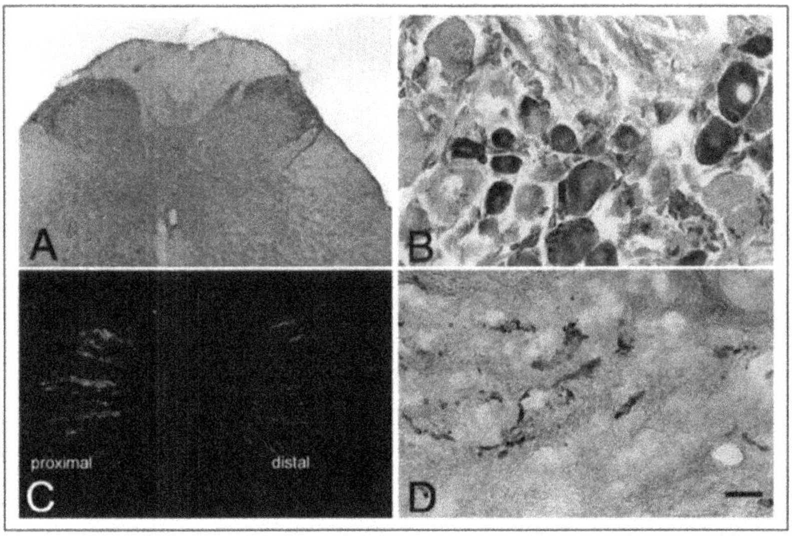

Fig. 1. Immunohistochemical (A, B, D) and immunofluorescence (C) staining of μ opioid receptors in rat dorsal horn of the spinal cord (A), dorsal root ganglia (B), ligated sciatic nerve ipsilateral to the inflamed hindpaw (C) and in inflamed subcutaneous paw tissue (D).

endogenous opioid peptides specific at μ receptors have been detected recently: endomorphin-1 and endomorphin-2.[18]

POMC-derived peptides were first reported in human lymphocytes by Blalock and Smith.[20] Since then, POMC-related opioid peptides have been found in immune cells of many vertebrates and nonvertebrates (reviewed in refs. 20, 21 and in chapters 4, 11). Extending the initial notion that only truncated forms of POMC messenger ribonucleic acid (mRNA) are present in immune cells,[20,22] full-length POMC mRNA, identical in sequence to that isolated from the pituitary gland, has been demonstrated in rat mononuclear leukocytes recently. This POMC transcript is spliced in the same way as the pituitary transcript and consequently contains the sequence for the signal peptide. The POMC protein is also proteolytically processed in a way consistent with the pituitary gland (see ref. 23 and chapter 4).

PENK-derived opioid peptides have also been detected in human and rodent immune cells (reviewed in ref. 24 and in chapter 4). Upon in vitro stimulation or under pathological conditions these cells express enhanced levels of PENK mRNA. In subpopulations of these cells this mRNA is highly homologous to brain PENK mRNA, abundant and apparently translated, since immunoreactive met-enkephalin is present and/or released.[24] PENK mRNA and enkephalins have been found in activated $CD4^+$ (activated/memory T cell marker) T cells,[25] $CD4^+$ thymocytes,[26] macrophages, monocytes and mast cells.[27] The appropriate enzymes necessary for posttranslational processing of both POMC and PENK have also been identified in immune cells (see ref. 28 and chapter 4).

Thus, a growing body of evidence indicates that both POMC- and PENK-derived opioid peptides are produced by immune cells. Immune-derived opioid peptides are involved in the modulation of inflammatory pain.[2] Apparently, persistent inflammation is a pathophysiological in vivo stimulus for the immune system and represents a condition that is closer to the clinical setting than some of the early in vitro studies. Recent studies by

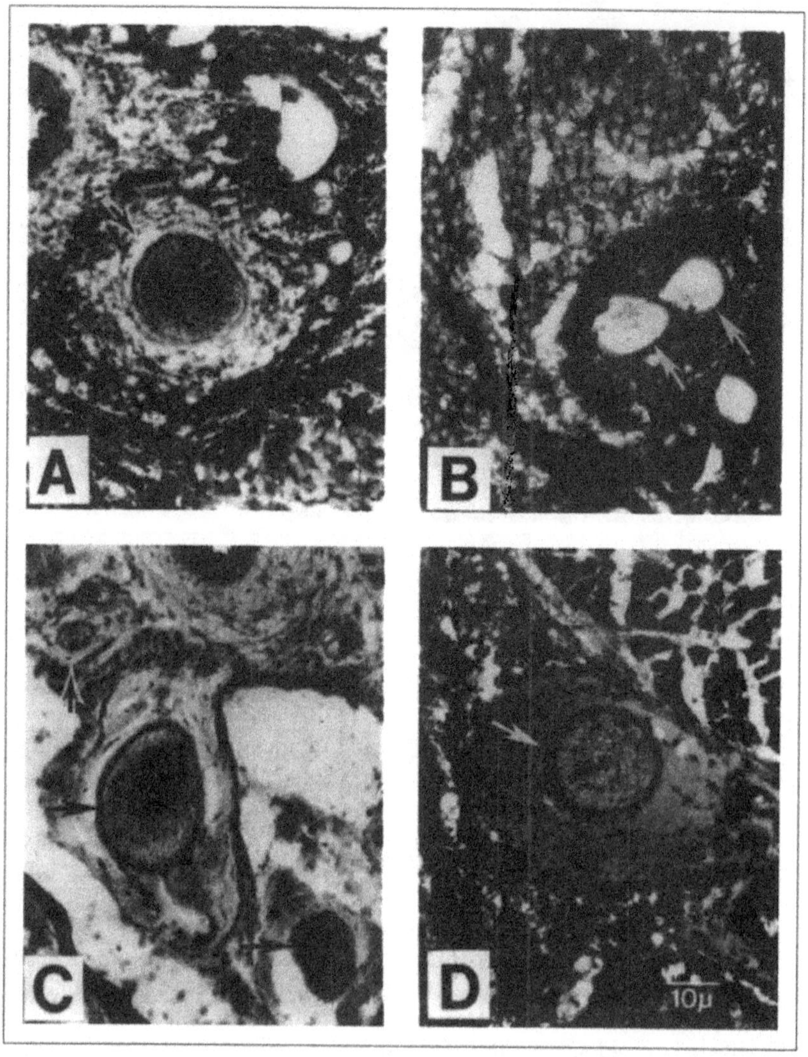

Fig. 2. Cross-sections of rat plantar subcutaneous paw tissue. After intraplantar injection of horseradish peroxidase, reaction products are seen in epineurial connective tissue and in perineurium of plantar nerves (A-D). The endoneurium of nerves is stained only in inflamed paws (A) and after intraplantar injection of mannitol, hypertonic sloution (C) or hypertonic saline (D), but not in noninflamed paws (B). Reprinted with permission from Antonijevic et al. J Neurosci 1995; 15(1):165-172. ©1995 Society for Neuroscience.

Mechanick et al[29] detected POMC mRNA in macrophages and monocytes by in situ hybridization. In addition, in double immunohistochemistry β-endorphin was localized in macrophages and monocytes of lung and spleen but not in lymphocytes. In CFA-induced arthritis levels of adrenocorticotrophin and β-endorphin are increased in the spleen and thymus in rats.[30] Also, in CFA-induced hind paw inflammation in rats, immune cells which have migrated to the inflamed tissue contain elevated levels of POMC, PENK and prodynorphin mRNA and of the corresponding opioid peptides β-endorphin, met-enkephalin

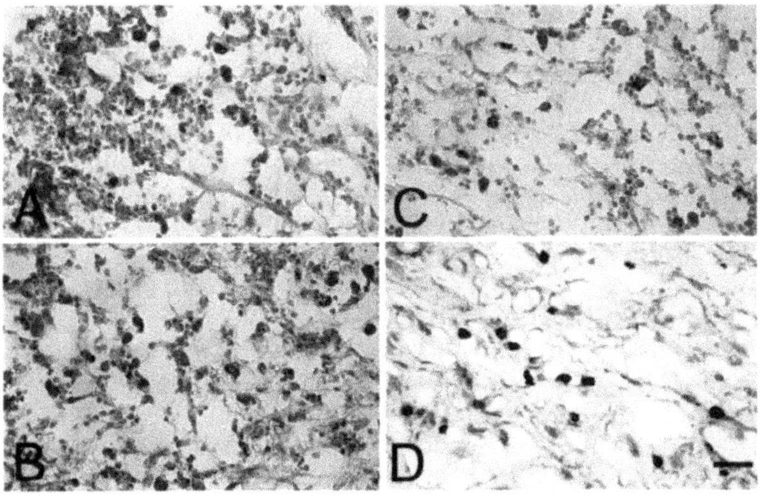

Fig. 3. Immunohistochemical localization of opioid peptides and corticotropin-releasing factor (CRF) in inflamed subcutaneous paw tissue in the rat. A, β-endorphin; B, met-enkephalin; C, dynorphin; D, CRF.

and dynorphin (Fig.3).[31] The detection of mRNAs encoding opioid peptide precursors in immune cells during inflammation suggests that these peptides are synthesized locally. Histomorphological and double-staining procedures have identified the β-endorphin containing cells as T- and B-lymphocytes as well as monocytes and macrophages.[31,32] Our recent studies using double immunofluorescence introduced more anatomical evidence that β-endorphin is present in activated/memory T cells. Indeed, we found that memory but not naive T cells are the predominant population in the inflamed subcutaneous tissue that contain β-endorphin.[6] This is consistent with the widely held view that activated/memory T cells migrate to peripheral inflamed tissue, while naive cells typically do not enter peripheral but migrate to lymphoid tissue.[33] Many studies have shown that immunocytes can produce and release β-endorphin and that this peptide is identical to its pituitary gland counterpart in terms of bioactivity, antigenicity, and molecular weight.[3,23] Our double staining experiments showed that β-endorphin and the activated/memory T cell marker CD4 were largely colocalized.[6] Also, some β-endorphin positive cells stained for the macrophage/monocyte marker ED1.[6] Recently, we showed that recruitment of more opioid peptide-containing immune cells led to more profound analgesia.[34] Accordingly, immunohistochemical procedures found a substantial increase of immune cells containing β-endorphin with the duration of inflammation (2 hours-4 days).[34]

Expression of Corticotropin-Releasing Factor and Interleukin-1 (IL-1) Receptors on Immune Cells

CRF and IL-1 are the major agents releasing opioids from immune cells (reviewed in chapters 3, 7). CRF is produced predominantly in the paraventricular nucleus of the hypothalamus and delivered into portal capillaries converging in the anterior lobe of the pituitary.[35] CRF is also widely expressed in extracranial tissues such as the immune system[36] but at levels much lower than in hypothalamus. In parallel to the increased CRF-IR in immune cells (lymphocytes and monocytes/macrophages; Fig. 4),[37] specific binding sites for radiolabeled CRF have been shown on resident macrophages of the mouse spleen,[38] on human monocytes, macrophages and T lymphocytes.[39] Using autoradiographic techniques we were able to detect receptors for CRF and IL-1 on monocytes/ macrophages and T lymphocytes in inflamed subcutaneous tissue (Fig. 4).[40]

These anatomical findings are consistent with the finding that CRF or IL-1 can induce the secretion of opioid peptides from immune cells.[41,42] Short-term incubation with CRF or IL-1 can release β-endorphin from immune cell suspensions prepared from lymph nodes in vitro (see ref. 3 and chapters 3, 7).

Expression of Adhesion Molecules

The mechanisms underlying the migration of opioid-containing immunocytes to inflamed tissue are beginning to be unraveled. The recruitment of immune cells to sites of inflammation is a multistep process involving the sequential activation of various adhesion molecules located on immune cells and vascular endothelium. Initially, members of the selectin family present on leukocytes (L-selectin) and endothelial cells (P- and E-selectin) tether flowing leucocytes to endothelium which subsequently roll along the blood vessel wall. In the second step, chemoattractants activating integrins and Ig superfamily members mediate firm adhesion to the endothelium. In the final step, chemotaxis and transmigration across the endothelial lining into the surrounding tissue take place, mediated by e.g., platelet-endothelial adhesion molecule-1 (PECAM-1) (see ref. 43 and chapter 7).

Using specific monoclonal antibodies against adhesion molecules we examined the expression of E-, P- and L-selectin and PECAM-1, and their co-expression with β-endorphin in inflamed and noninflamed subcutaneous tissue and lymph nodes.[44] We showed that L-selectin was expressed on immune cells (lymphocytes and macrophages) in lymph nodes and on cells migrating to the inflamed subcutaneous paw tissue. P-selectin and PECAM-1 were constitutively expresed on endothelium of blood vessels in noninflamed lymph nodes and subcutaneous paw tissue and they were up-regulated in inflammation. Double immunofluorescence demonstrated that β-endorphin-positive cells stained for L-selectin in lymph nodes and inflamed subcutaneous paw tissue (Figs. 5). The β-endorphin positive cells expressing L-selectin appeared to be more abundant in inflamed compared to noninflamed tissue. Double staining experiments revealed that P-selectin and PECAM-1 were present on endothelial cells but not on β-endorphin-positive cells in lymph nodes and subcutaneous paw tissue (Figs. 6). These findings suggest the requirement of PECAM-1, L- and P-selectin for the migration of immunocytes containing β-endorphin to peripheral inflamed tissue and indicate the involvement of adhesive mechanisms in pain control.[44] Indeed, recently we showed that selectin blockade strongly reduced endogenous peripheral opioid analgesia (see ref. 1 and chapter 7).

Clinical Studies

Clinical trials have demonstrated the analgesic efficacy of endogenous opioids and of small, systemically inactive doses of exogenous opioid receptor agonists administered into the vicinity of peripheral nerve teminals (see refs. 45-48 and chapters 5, 7). Tissue samples from inflamed human synovia exhibited specific binding of [3H]naloxone, indicating the presence of of opioid binding sites/receptors.[45] Moreover, opioid receptors were identified on peripheral terminals of sensory neurons in inflamed synovial tissue from patients un-

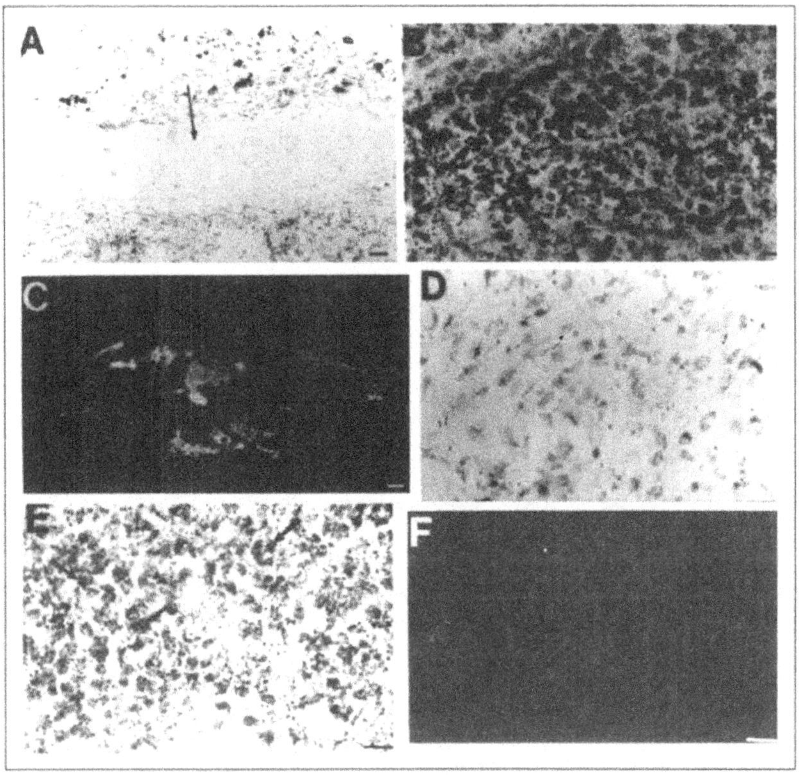

Fig. 4. Photoemulsion autoradiography of ^{125}I-corticotropin releasing factor (CRF)- (A, B) and ^{125}I-interleukin (IL)-1β- (D, E) binding sites in noninflamed (A, D) and inflamed (B, E) subcutaneous paw tissue in the rat. (F) shows displacement of ^{125}I-IL-1β binding by unlabeled IL-1β in inflamed tissue. Note binding on lymphocytes (large arrow) and macrophages (small arrow) in (B, E). Darkfield photomicrographs of tritium-sensitive ultrafilm show the distribution of ^{125}I-CRF binding in the inflammatory foci (arrow) of subcutaneous tissue (C) and nonspecific binding in the presence of 1 μM unlabeled CRF (F).

dergoing arthroscopic knee surgery using receptor autoradiography.[47] The presence of β-endorphin was immunohistologically demonstrated in synovial tissue biopsied from patients with rheumatoid arthritis and osteoarthritis.[49] The amount of β-endorphin in culture supernatants of synovial tissue explants was also determined by radioimmunoassay.[49] Morphological studies revealed the presence of β-endorphin and met-enkephalin in lymphocytes, macrophages and mast cells in inflamed human synovia (Fig. 7).[47]

Summary

The immune system is a source of opioid peptides and plays an important role in the control of inflammatory pain. Inflammation not only increases the opioid receptor expression in DRG neurons but also enhances transport and accumulation of opioid receptors on the peripheral terminals of sensory neurons. Immune cells containing opioid peptides migrate to the inflamed tissue. This is orchestrated by adhesion molecules up-regulated on vessel endothelia and co-expressed by opioid-containing immunocytes. The peptides are secreted by stressful stimuli, CRF and cytokines and the corresponding receptors are

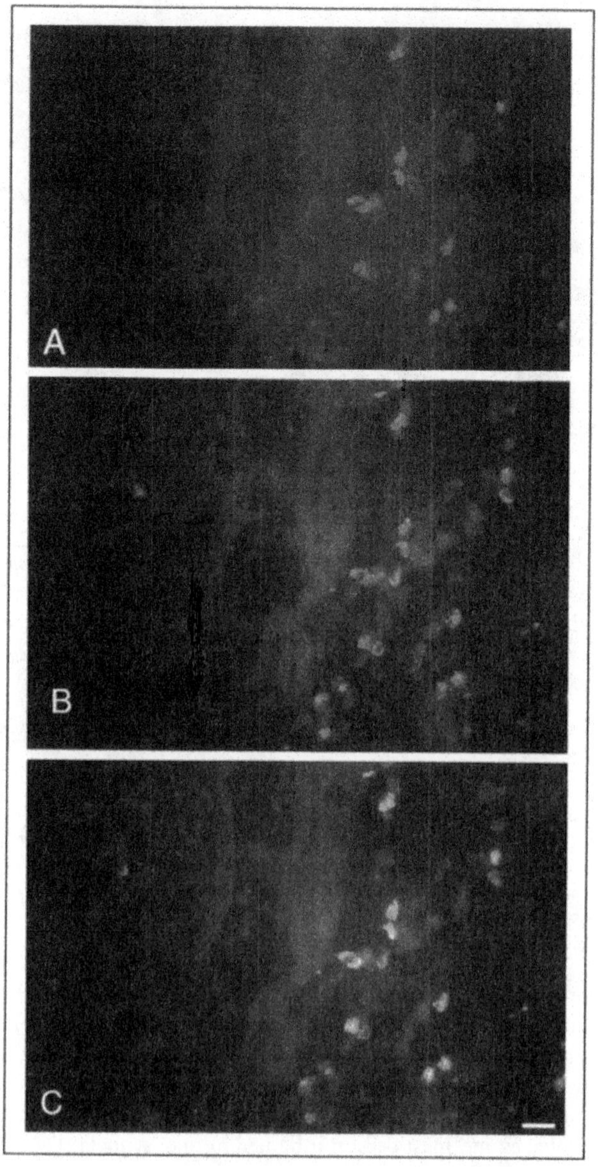

Fig. 5. Colocalization of β-endorphin and L-selectin in subcutaneous paw tissue in the rat. β-endorphin (red; A, B, C); L-selectin (green; A, B); L-selectin/β-endorphin (yellow; C). Bar = 20 μm. Reprinted with permission from Mousa et al. J Neuroimmunol 2000; 108:160-170. ©2000 Elsevier Science.

present on opioid-expressing leukocytes. The opioids bind to their receptors localized on peripheral sensory nerves leading to pain inhibion. In the more distant future, these findings might stimulate the development of novel analgesics based on enhancing the transport and release of immune-derived opioid peptides into injured tissue.

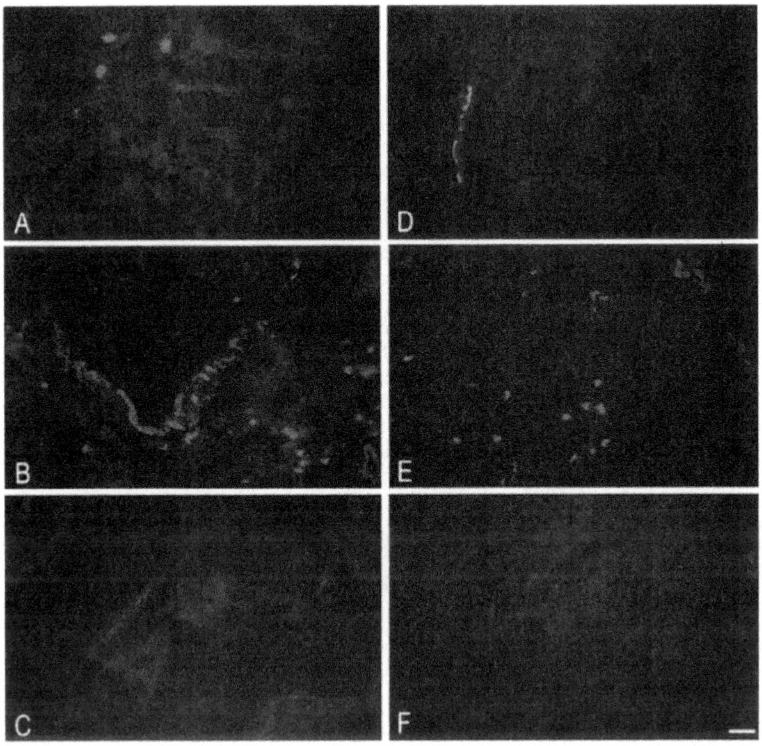

Fig. 6. Double-staining of β-endorphin (red) with P-selectin (green) (A, B) and with PECAM-1 (green) (D, E) in noninflamed (A, D) and inflamed (B, E) subcutaneous paw tissue in the rat. C and F are controls: Preabsorption of β-endorphin antiserum with β-endorphin or the omission of the primary antibodies for P-selectin (C) or PECAM (F) revealed no significant immunoreactivity. Bar = 20 μm. Reprinted with permission from Mousa et al J Neuroimmunol 2000; 108:160-170. ©2000 Elsevier Science.

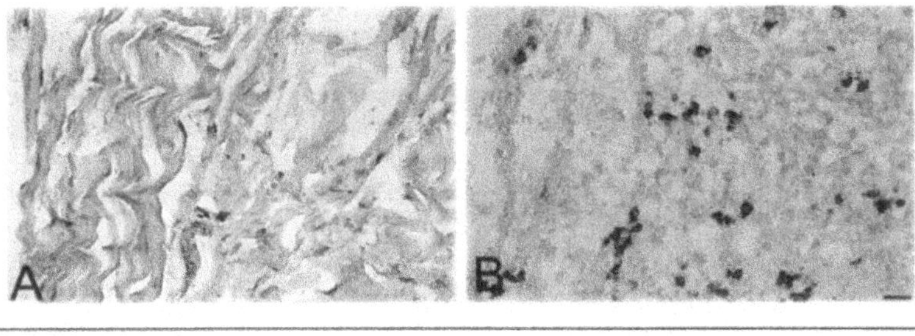

Fig. 7. Immunohistochemical localization of β-endorphin in human synovium. A, noninflamed synovium; B, inflamed synovium. Bar = 20 μm.

References

1. Machelska H, Cabot PJ, Mousa SA et al. Pain control in inflammation governed by selectins. Nat Med 1998; 4:1425-28.
2. Stein C. Mechanisms of Disease: The Control of Pain in Peripheral Tissue by Opioids. N Engl J Med 1995; 332:1685-90.
3. Cabot PJ, Carter L, Gaiddon C et al. Immune cell-derived β-endorphin: production, release and control of inflammatory pain in rats. J Clin Invest 1997; 100:142-148.
4. Stein C, Millan MJ, Shippenberg TS et al. Peripheral opioid receptors mediating antinociception in inflammation Evidence for involvement of mu, delta and kappa receptors. J Pharmacol Exp Ther 1989; 248:1269-75.
5. Zhang Q, Schaffer M, Elde R et al. Effects of neurotoxins and hindpaw inflammation on opioid receptor immunoreactivities in dorsal root ganglia. Neuroscience 1998; 85:281-91.
6. Mousa SA, Zhang Q, Ji RR. et al. β-Endorphin containing memory-cells and μ-opioid receptors undergo site-directed transport into peripheral inflamed tissue. J Neuroimmunol 2001; 115(1-2):71-8.
7. Ji R-R, Zhang Q, Law P-Y et al. Expression of μ-, δ-, and κ-opioid receptor-like immunoreactivities in rat dorsal root ganglia after carrageenan-induced inflammation. J Neurosci 1995; 15:8156-8166.
8. Zhou L, Zhang Q, Stein C et al. Contribution of opioid receptors on primary afferent versus sympathetic neurons to peripheral opioid analgesia. J Pharmacol Exp Ther 1998; 261:1-7.
9. Laduron PM. Axonal transport of opiate receptors in capsaicin-sensitive neurones. Brain Res 1984; 294(1):157-60.
10. Hassan AHS, Ableitner A, Stein C etal. Inflammation of the rat paw enhances axonal transport of opioid receptors in the sciatic nerve and increases their density in the inflamed tissue. Neurosci 1993; 55:185-195.
11. Stein C, Hassan AHS, Przewlocki R etal. Opioids from immunocytes interact with receptors on sensory nerves to inhibit nociception in inflammation. Proc Natl Acad Sci USA 1990; 87:5935-5939.
12. Antonijevic I, Mousa SA, Schafer M et al. Perineurial defect and peripheral opioid analgesia in inflammation. J Neurosci 1995; 15:165-72.
13. Coggeshall RE, Zhou S, Carlton SM. Opioid receptors on peripheral sensory axons. Brain Res 1997; 764:126-132.
14. Olsson Y. Microenvironment of the peripheral nervous system under normal and pathological conditions. Crit Rev Neurobiol 1990;5:265-311.
15. de la Motte DJ, Hall SM, Allt G. A study of the perineurium in peripheral nerve pathology. Acta Neuropathol (Berl) 1975; 33:257-70.
16. Rechthand E, Rapoport SI. Regulation of the microenvironment of peripheral nerve: role of the blood-nerve barrier. Prog Neurobiol 1987; 28:303-43.
17. Höllt V. Opioid peptide processing and receptor selectivity. Annu Rev Pharmacol Toxicol 1986; 26:59-77.
18. Zadina JE, Hackler L, Ge L-J, Kastin AJ. A potent and selective endogenous agonist for the μ-opiate receptor. Nature 1997; 386:499-502.
19. Blalock JE, Smith EM. Human leukocyte interferon: structural and biological relatedness to adrenocorticotropic hormone and endorphins. Proc Natl Acad Sci USA 1980; 77:5972-4.
20. Panerai AE, Sacerdote P. β-endorphin in the immune system: a role at last? Immunology Today 1997; 18:317-319.
21. Blalock JE. The syntax of immune-neuroendocrine communication. Immunol Today 1994; 15:504-511.
22. Sharp B, Yaksh T. Pain killers of the immune system. T lymphocyte produce opioid immunopeptides that control pain at sites of inflammation. Nature Med 1997; 3:831-832.
23. Lyons PD, Blalock JE. Pro-opioimelanocortin gene expression and protein processing in rat mononuclear leukocytes. J Neuroimmunol 1997; 78:47-56.
24. Weisinger G. The transcriptional regulation of the preproenkephalin gene. Biochem J 1995; 307:617-629.
25. Zurawski G, Benedik M, Kamb BJ et al. Activation of mouse T-helper cells induces abundant preproenkephalin mRNA synthesis. Science 1986; 232:772-5.
26. Linner KM, Nicol SE, Sharp BM. IL-1 beta modulates the concanavalin-A-induced expression of proenkephalin A mRNA in murine thymocytes. J Pharmacol Exp Ther 1993; 267(3):1566-72.
27. Martin J, Prystowsky MB, Angeletti RH. Preproenkephalin mRNA in T-cells, macrophages, and mast cells. J Neurosci Res 1987; 18:82-7.
28. Vindrola O, Mayer AMS, Citera G et al. Prohormone convertases PC2 and PC3 in rat neutrophils and macrophages. Neuropeptides 1994; 27:235-244.

29. Mechanick JI, Levin N, Roberts JL et al. Proopiomelanocortin gene expression in a distinct population of rat spleen and lung leukocytes. Endocrinology 1992; 131:518-25.
30. Jessop DS, Renshaw D, Lightman SL et al. Changes in ACTH and beta-endorphin immunoreactivity in immune tissues during a chronic inflammatory stress are not correlated with changes in corticotropin-releasing hormone and arginine vasopressin. J. Neuroimmunol. 1995; 60:29-35.
31. Przewlocki R, Hassan AHS, Lason W et al. Gene expression and localization of opioid peptides in immune cells of inflamed tissue. Functional role in antinociception. Neuroscience 1992; 48:491-500.
32. Hassan AHS, Przewlocki R, Herz A et al. Dynorphin, a preferential ligand for kappa-opioid receptors, is present in nerve fibers and immune cells within inflamed tissue of the rat. Neurosci Lett 1992; 140:85-88.
33. Westermann J, Pabst R. How organ-specific is the migration of 'naive' and 'memory' T cells? Immunol Today 1996; 17:278-82.
34. Rittner HL, Brack A, Machelska H et al. Opioid peptides expressing leukocytes –identification, recruitment and simultaneously increasing inhibition of inflammatory pain. Anesthesiology 2001, In press.
35. Aguilera G. Corticotropin releasing hormone, receptor regulation and the stress response. Trends Endocrinol. 1998; 9:329-336.
36. Brouxhon SM, Prasad AV, Joseph SA et al. Localization of corticotropin-releasing factor in primary and secondary lymphoid organs of the rat. Brain Behav Immun 1998; 12(2):107-22.
37. Schäfer M, Mousa SA, Zhang Q et al. Expression of corticotropin-releasing factor in inflamed tissue is required for intrinsic peripheral opioid analgesia. Proc Natl Acad Sci USA 1996; 93:6096-6100.
38. Webster EL, Tracey DE, Jutila MA et al. Corticotropin-releasing factor receptors in mouse spleen: identification of receptor-bearing cells as resident macrophages. Endocrinology 1990; 127:440-52.
39. Audhya T, Jain R, Hollander CS. Receptor-mediated immunomodulation by corticotropin-releasing factor. Cell Immunol 1991; 134:77-84.
40. Mousa SA, Schafer M, Mitchell WM et al. Local upregulation of corticotropin-releasing hormone and interleukin-1 receptors in rats with painful hindlimb inflammation. Eur J Pharmacol 1996; 311:221-31.
41. Schäfer M, Carter L, Stein C. Interleukin-1β and corticotropin-releasing-factor inhibit pain by releasing opioids from immune cells in inflamed tissue. Proc Natl Acad Sci USA 1994; 91:4219-4223.
42. Kavelaars A, Berkenbosch F, Croiset G et al. Induction of beta-endorphin secretion by lymphocytes after subcutaneous administration of corticotropin-releasing factor. Endocrinology 1990; 126:759-64.
43. Butcher EC, Picker LJ. Lymphocyte homing homeostasis. Science 1996; 272:60-66.
44. Mousa SA, Machelska H, Schafer M et al. Co-expression of beta-endorphin with adhesion molecules in a model of inflammatory pain. J Neuroimmunol 2000; 108(1-2):160-70.
45. Lawrence AJ, Joshi GP, Michalkiewicz A et al. Evidence for analgesia mediated by peripheral opioid receptors in inflamed synovial tissue. Eur J Clin Pharmacol 1992; 43(4):351-5.
46. Stein C, Hassan AH, Lehrberger K et al. Local analgesic effect of endogenous opioid peptides. Lancet 1993; 342(8867):321-4.
47. Stein C, Pfluger M, Yassouridis A et al. No tolerance to peripheral morphine analgesia in presence of opioid expression in inflamed synovia. J Clin Invest 1996; 98(3):793-9.
48. Stein A, Yassouridis A, Szopko C et al. Intraarticular morphine versus dexamethasone in chronic arthritis. Pain 1999; 83(3):525-32.
49. Koiwa M, Shiga H, Nakamura H et al. Yoshino S. Role of opioid peptide in rheumatoid arthritis—detection of beta-endorphin in synovial tissue]. Arerugi 1992; 41(9):1423-9.

CHAPTER 7

Functional Evidence of Pain Control by the Immune System

Halina Machelska

Introduction

Pain can be effectively controlled by various endogenous mechanisms. Recent research has shown that these mechanisms are not restricted to the central nervous system. Intrinsic pain inhibition can occur also in the periphery, mediated by an interaction between immune cells and sensory nerve endings. A prerequisite for the manifestation of such peripheral effects seems to be inflammation, accompanied by hyperalgesia. Opioid receptors are present on peripheral endings of sensory nerves and are up-regulated during the development of inflammation. Their endogenous ligands, opioid peptides, are synthesized in circulating immune cells which, under pathological conditions, migrate to injured sites. This is orchestrated by selectins, adhesion molecules located on opioid-containing immune cells and on vascular endothelium. Under environmental stressful stimuli or in response to releasing agents (e.g., corticotropin releasing factor, CRF; cytokines) these immunocytes can secrete opioids. These activate peripheral opioid receptors and produce antinociception (analgesia) by inhibiting either the excitability of the nerves or the release of excitatory, proinflammatory neuropeptides. These effects can be abolished by opioid receptor antagonists, antibodies against opioid peptides, by immunosuppression and by blocking the selectin-dependent extravasation of opioid-containing immune cells.

Peripheral Opioid Receptors

All three opioid receptors (μ, δ and κ) are expressed within sensory neurons. They have been found on cell bodies in the dorsal root ganglia and on central and peripheral terminals of primary afferent neurons in animals and in humans. This was confirmed by in vivo experiments which showed that the peripheral analgesic effects of μ-, δ- and κ-selective agonists are abolished by pretreatment with capsaicin, a neurotoxin selective for primary afferent neurons. In inflammation, the number of opioid receptors increases on peripheral nerve terminals. In addition, pre-existent, but possibly inactive neuronal opioid receptors can become active owing to the specific milieu (e.g., low pH) of inflamed tissue. Furthermore, inflammation entails a disruption of the perineurium (a normally rather impermeable barrier encasing peripheral nerve fibers) and increases the number of peripheral sensory nerve terminals in inflamed tissue (reviewed in ref. 1 and in the chapter by Stein in this volume). Activation of these opioid receptors results in potent peripherally

Immune Mechanisms of Pain and Analgesia, edited by Halina Machelska and Christoph Stein.

mediated analgesia, particularly in inflamed subcutaneous tissue, viscera and joints, and confers anti-inflammatory effects (reviewed in refs. 2-6, in "Peripheral analgesic effects of immune cell-derived opioid peptides" in this chapter and in chapters by Stein and Walker in this volume).

Peripheral Opioid Peptides

Opioid peptides are the natural ligands at opioid receptors. Three families of these peptides are well characterized in the central nervous and neuroendocrine systems. Each family derives from a distinct gene and precursor protein: proopiomelanocortin, proenkephalin and prodynorphin. Appropriate processing yields their respective major representative opioid peptides β-endorphin, enkephalin and dynorphin. Each peptide exhibits different affinity and selectivity for the three receptor types μ (β-endorphin, enkephalin), δ (enkephalin, β-endorphin) and κ (dynorphin).[7] Recently, two additional endogenous opioid peptides have been isolated from bovine brain: endomorphin-1 and endomorphin-2. Their precursors are not known yet. These two peptides are considered to have the highest specificity and affinity for μ-receptors of any endogenous substance so far described.[8] Proopiomelanocortin- and proenkephalin-derived opioid peptides have been detected in rodent and human immune cells[9-12] (reviewed in the chapter by Smith in this volume).

Immune-derived opioid peptides apparently play a substantial role in the modulation of inflammatory pain.[4] In inflamed rat paw mRNAs encoding proopiomelanocortin and proenkephalin and their respective opioid peptide products β-endorphin and enkephalin are found predominantly within T- and B-lymphocytes as well as in monocytes, macrophages and granulocytes.[13-16] Small amounts of dynorphin are also detectable by immunocytochemistry[17] (reviewed in the chapter by Mousa in this volume).

Other sources of opioids in the periphery are the adrenals and the pituitary, but these have been excluded as sources for opioid ligands at peripheral receptors. Opioid peptides have been found in sensory ganglia and in peripheral terminals of sensory nerves (reviewed in ref. 18) but direct functional evidence regarding their possible role in peripheral nociceptive transmission is to date lacking.

Interactions of Immune-Derived Opioids with Peripheral Opioid Receptors

Migration of Opioid-Containing Immune Cells to Inflamed Tissue

The recruitment of leukocytes into areas of inflammation begins with the binding of white blood cells to endothelium, followed by their transmigration into tissues. Although this observation has been documented for more than 150 years, only the last decade, with the identification of specific cell adhesion and chemoattractant/activator molecules (Fig. 1) uncovered the molecular mechanisms underlying leukocyte extravasation.

Leukocytes are recruited from the circulation by a well-orchestrated set of events (Fig. 2). In these processes, leukocytes undergo multiple attachments to, and detachments from, the vessel endothelial cells, prior to transendothelial cell migration. This includes slowing and rolling along the endothelial cell wall that is mediated predominantly by the interaction of selectins expressed on leukocytes (L-selectin) and endothelial cells (P- and E-selectin) with their ligands expressed on endothelium or immune cells, respectively. The rolling immunocytes can then be activated by chemoattractants released from inflammatory cells

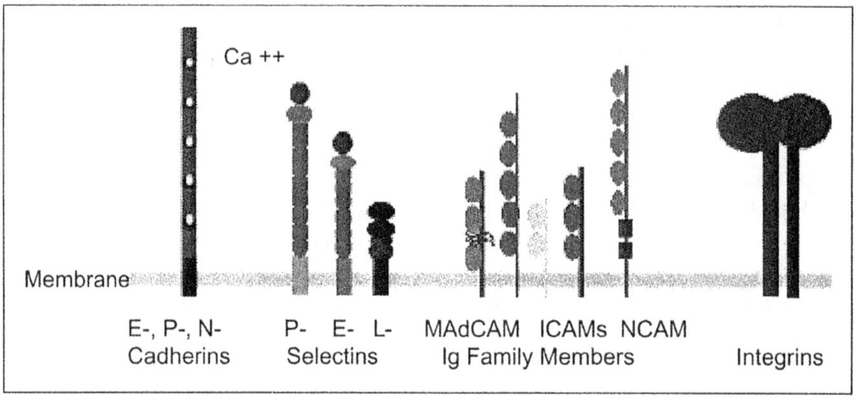

Fig. 1. Schematic structure of adhesion molecules. The cadherins are transmembrane proteins with five motifs in the extracellular domain which are bridged by Ca^{++}. The selectins are transmembrane proteins that contain an N-terminal lectin domain, epidermal growth factor-type domain and two to nine complement regulatory repeats. The Ig superfamily members are structurally most diverse, but each have two to five Ig repeats in the extracellular domain and most of them are transmembrane proteins. The integrins are noncovalently linked heterodimers composed of α and β subunits. MAdCAM, mucosal-addressin adhesion molecule; ICAMs, intercellular adhesion molecules; NCAM, neuronal adhesion molecule. Reprinted from Am J Med 1999, 106, Petruzzelli L, Takami M, Humes D. Structure and function of cell adhesion molecules. 467-476, Copyright 1999, with permission from Experta Medica Inc.

and endothelium. This leads to up-regulation and increased avidity of integrins. These mediate the firm adhesion of leukocytes to endothelial cells via ligands of the immunoglobulin (Ig) superfamily. Finally, the immune cells transmigrate through the endothelial wall mediated by Ig superfamily members (e.g., platelet-endothelial adhesion molecule-1; PECAM-1) and are directed to the sites of inflammation to initiate a host defense (reviewed in refs. 19-21).

Recent findings suggest that these events can also be involved in endogenous control of inflammatory pain. P-selectin and PECAM-1 present on endothelia are up-regulated in inflammation. The number of immunocytes expressing L-selectin is increased in inflamed subcutaneous paw tissue. Most importantly, many β-endorphin-containing cells express L-selectin in inflamed tissue[22] (for details see the chapter by Mousa in this volume). Furthermore, pretreatment of rats with a selectin blocker (fucoidin) decreases the number of β-endorphin-containing immunocytes infiltrating the inflamed site.[23] In consequence, this diminishes the β-endorphin content in inflamed tissue and concurrently abolishes peripheral opioid analgesia (for details see ref. 23 and "Peripheral analgesic effects of immune cell-derived opioid peptides" in this chapter). This suggests that circulating β-endorphin-producing immunocytes home to inflamed tissue where they secrete the opioid to inhibit pain. Afterwards they travel to the regional lymph nodes, depleted of the opioid peptide.[14] This migratory pattern is reminiscent of memory cells. The trafficking of those cells is not random but they are specifically directed to sites of antigenic or microbial invasion (e.g., inflammatory lesions of the skin).[19,20] Consistent with this notion, β-endorphin was indeed found in memory-type T-cells (the chapter by Mousa in this volume).[14,15] These findings suggest that local signals not only stimulate the synthesis of opioid peptides in resident inflammatory cells but also attract opioid-containing cells from the circulation to the site of tissue injury to reduce pain (for details see "Peripheral analgesic effects of immune cell-derived opioid peptides" in this chapter).

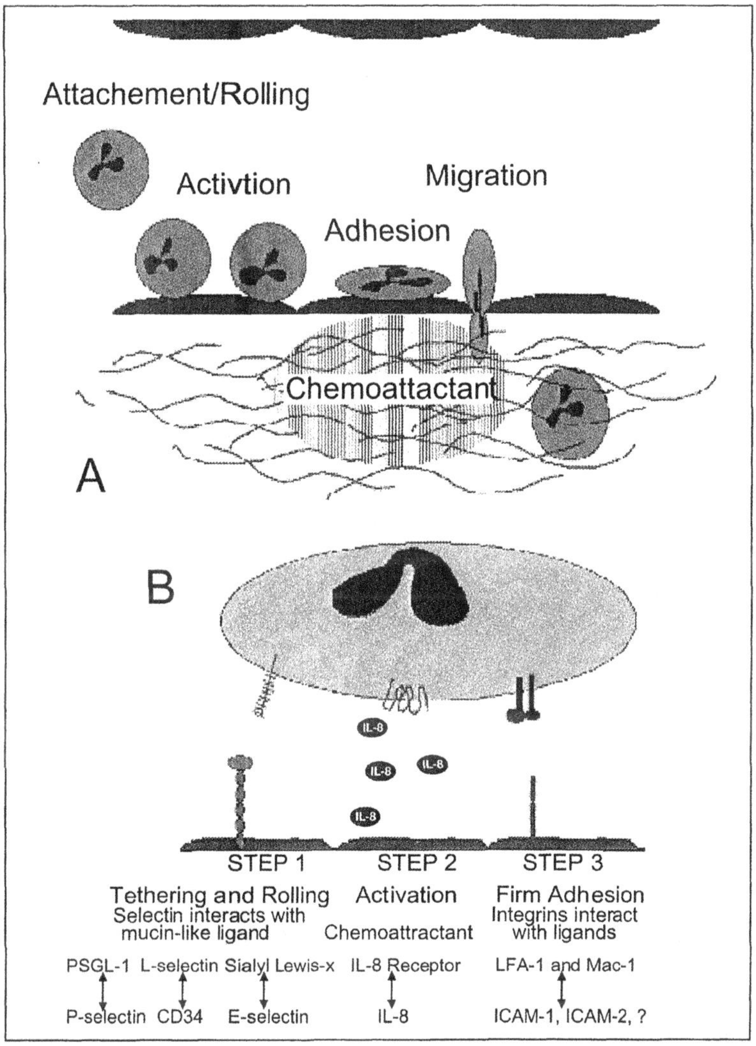

Fig. 2. The recruitment of leukocytes from circulation to the site of inflammation. As seen in A, recruitment of leukocytes consists of four major steps: attachement and rolling, activation, adhesion and migration. Specific molecular events play a role at each step, and those are depicted in B. For details see text. PSLG-1, P-selectin glycoprotein ligand-1; integrins: LFA-1, lymphocyte function-associated antigen-1 (CD11a/CD18); Mac-1 (CD11b/CD18); Ig superfamily members: ICAM (intercellular adhesion molecule)-1, -2 . Reprinted from Am J Med 1999, 106, Petruzzelli L, Takami M, Humes D. Structure and function of cell adhesion molecules. 467-476, Copyright 1999, with permission from Experta Medica Inc.

Release of Opioid Peptides from Immune Cells

Once the immune cells reach the site of inflammation, they have to secrete the opioids to produce pain relief. Endogenous agents triggering opioid release within inflamed tissue are CRF and cytokines. CRF mRNA and immunoreactivity have been demonstrated in various lymphoid tissues such as in thymus, spleen and in T and B lymphocytes. In inflamed subcutaneous tissue CRF is detectable in immune cells, fibroblasts and vascular

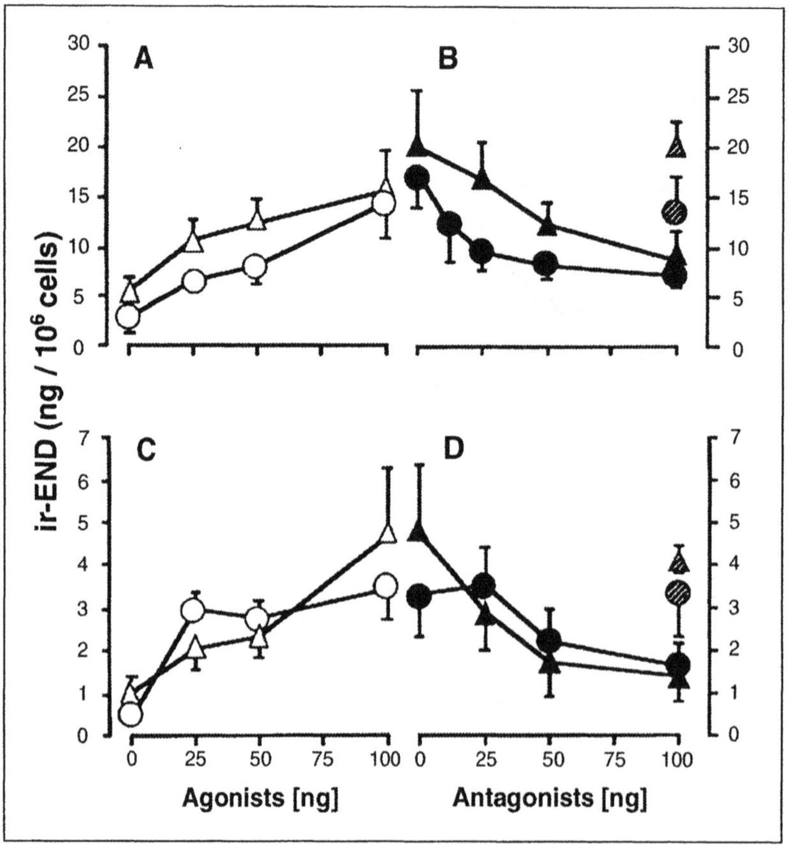

Fig. 3. IL-1- and CRF-induced release of β-endorphin (A, C) and antagonism by IL-1ra and α-helical-CRF (B, D) in cell suspensions from noninflamed (A, B) and inflamed (C, D) popliteal lymph nodes. Open circles: IL-1; open triangles: CRF; filled circles, IL-1ra (IL-1 receptor antagonist); filled triangles: α-helical-CRF (CRF receptor antagonist). IL-1 and CRF dose-dependently release β-endorphin. This was dose-dependently inhibited by their respective antagonists ($p<0.05$, regression ANOVA). The IL-1 (100 ng) effect was not significantly changed by α-helical-CRF (hatched circles) and the CRF (100 ng) effect was not significantly changed by IL-1ra ($p>0.05$, Mann-Whitney U test). Data are expressed as means ± SEM. IL-1, interleukin-1; CRF,corticotropin releasing factor. Reprinted with permission of the Journal of Clinical Investigation via the Copyright Clearance Center from Cabot JP et al. J Clin Invest 1997; 100:142-148.

endothelium. Interestingly, peripheral CRF expression is enhanced in inflamed synovial and subcutaneous tissue of animals and humans. CRF and interleukin-1β (IL-1) receptors have been detected in rat synovial and subcutaneous tissue on lymphocytes and monocytes/macrophages but not on peripheral sensory nerves. The number of both receptors is greatly enhanced in inflammation. Their pharmacolgical characteristics are similar to the high-affinity CRF and IL-1 binding sites in the pituitary (reviewed in ref. 24 and in the chapters by Schäfer and Mousa in this volume). Both CRF and IL-1 can release β-endorphin from immune cell suspensions prepared from lymph nodes in vitro (Fig. 3).[14,25] This release is specific to CRF and IL-1 receptors because it is reversed dose-dependently by their respective antagonists (Fig. 3), it is calcium-dependent and it is mimicked by elevated extracellular concentrations of potassium.[14] This is consistent with a regulated pathway of release from secretory vesicles, as in neurons and endocrine cells. In summary, these findings indicate that CRF and IL-1 can cause secretion of opioids from immune cells. These opioid peptides subsequently activate opioid receptors on sensory nerves to inhibit pain (for details see "Peripheral analgesic effects of immune cell-derived opioid peptides" in this chapter).

Peripheral Analgesic Effects of Immune Cell-Derived Opioid Peptides

In rat unilateral hind paw inflammation induced by complete Freund's adjuvant[26] endogenous opioid analgesia can be elicited by environmental stress (cold water swim; CWS).[27] In this model rats swim for 1 min in water of 4°C and pain thresholds are measured afterwards in a paw pressure test (Randall-Selitto test) evaluating a response to painful mechanical stimuli.[27] CWS produces analgesia in inflamed but not in noninflamed paws.[27] This effect is opioid specific because it is blocked dose-dependently and stereospecifically by the opioid receptor antagonist naloxone injected into the paw.[27] Using selective antagonists it was shown that μ and δ, but not κ receptors play a major role (Fig. 4).[27] A prominent opioid peptide involved in peripheral pain control is β-endorphin because CWS-induced analgesia is dose-dependently abolished by antibodies against β-endorphin but not against met-enkephalin or dynorphin (Fig. 5).[27,28] In parallel, exogenous application of β-endorphin into the inflamed paw produces dose-dependent analgesia reversed by μ- and δ-receptor antagonists.[27] The systemic (subcutaneous or intravenous) administration of opioid receptor antagonists or antibodies against opioid peptides does not change CWS-induced analgesia, demonstrating a peripheral site of action.[27,28] Thus, analgesia in inflamed tissue can by induced through the activation of local opioid receptors by endogenous opioid peptides (mainly β-endorphin) released during CWS.

An important stimulus for peripheral β-endorphin release appears to be CRF because CWS-induced analgesia is abolished by the local injection of a CRF receptor antagonist or antibody.[29] Consistently, the local application of small, systemically inactive doses of CRF produces potent CRF- and opioid receptor specific analgesic effects in inflamed, but not in noninflamed tissue.[25] Although exogenous IL-1 can release opioids from immune cells[14,25] (Fig. 3) and produce analgesia[25] endogenous IL-1 does not seem to be involved at this stage of inflammation (4 days)[29] (for details on the involvement of cytokines in peripheral opioid analgesia see the chapter by Schäfer in this volume). That immune cells are a source of opioids is demonstrated by abolishing CWS- and CRF-induced analgesia by immunosuppression with cyclosporine or whole body irradiation.[13,25,28] Moreover, these effects are also extinguished by blocking the extravasation of β-endorphin-containing immune cells. Fucoidin, an L- and P-selectin blocker can dose-dependently inhibit CWS- and CRF- but not fentanyl (an opioid receptor agonist)-induced analgesia in inflammation (Fig. 6).[23] Concurrently, the number of β-endorphin-containing cells and the total amount of β-endorphin in the tissue is significantly diminished.[23] Together, these findings demonstrate that L- and P-selectins regulate the migration of β-endorphin-containing immune cells and the subsequent generation of endogenous pain control in injured tissue.

Recently, the involvement of subpopulations of opioid-containing immunocytes and their contribution to endogenous analgesia was examined in relation to the development of inflammation.[16] In early (2 – 6 hours) inflammation a majority of opioid-producing leukocytes are granulocytes whereas at later stages (4 days) monocytes and macrophages play a dominant role. With increasing duration of inflammation the number of opioid-containing immunocytes and the β-endorphin content rise. In parallel, the

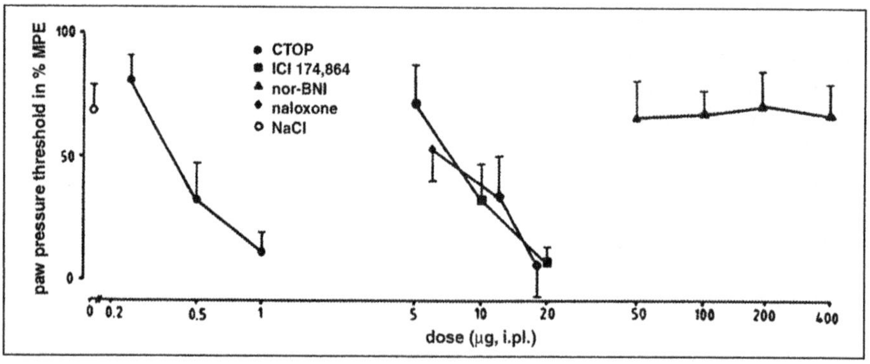

Fig. 4. Effects of local injections of naloxone and selective antagonists at µ- (CTOP), δ- (ICI 174,864) and κ- (nor-BNI) opioid receptors on CWS-induced analgesia in inflamed paws. Naloxone, CTOP and ICI 174,864, but not nor-BNI dose-dependently inhibit CWS-induced analgesia ($p<0.05$ and $p>0.05$, regression ANOVA, respectively). Data are expressed as means ± SEM. CTOP, D-Phe-Cys-Tyr-D-Trp-Orn-Thr-Pen-Thr-NH_2; ICI 174,864, N, N-diallyl-Thr-Aib-Aib-Phe-Leu; nor-BNI, nor-binaltorphimine. Reprinted with permission from Stein C et al. J Neurosci 1990; 10: 1292-1298. Copyright 1990 by the Society for Neuroscience.

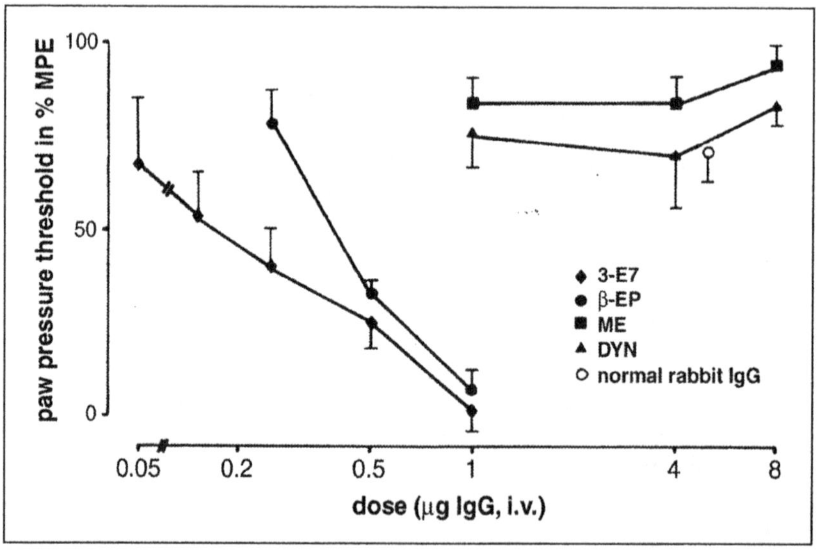

Fig. 5. Effects of local injections of antibodies against all opioids (3E7), β-endorphin (β-EP), Met-enkephalin (ME) or dynorphin A 1-17 (DYN) on CWS-induced analgesia in inflamed paws. 3E7 and anti-β-EP, but not anti-ME and anti-DYN dose-dependently inhibit CWS-induced analgesia ($p<0.05$ and $p>0.05$, regression ANOVA, respectively). Data are expressed as means ± SEM. Reprinted with permission from Stein C et al J Neurosci 1990; 10: 1292-1298. Copyright 1990 by the Society for Neuroscience.

CWS-induced analgesic effect increases.[16] Thus, the potency of endogenous pain inhibition is proportional to the number of opioid peptide-producing cells and distinct leukocyte lineages contribute to this function at different stages of inflammation. Further details on the neural mechanisms underlying the generation of peripheral opioid analgesia are discussed in ref. 18 and in the chapter by Stein in this volume.

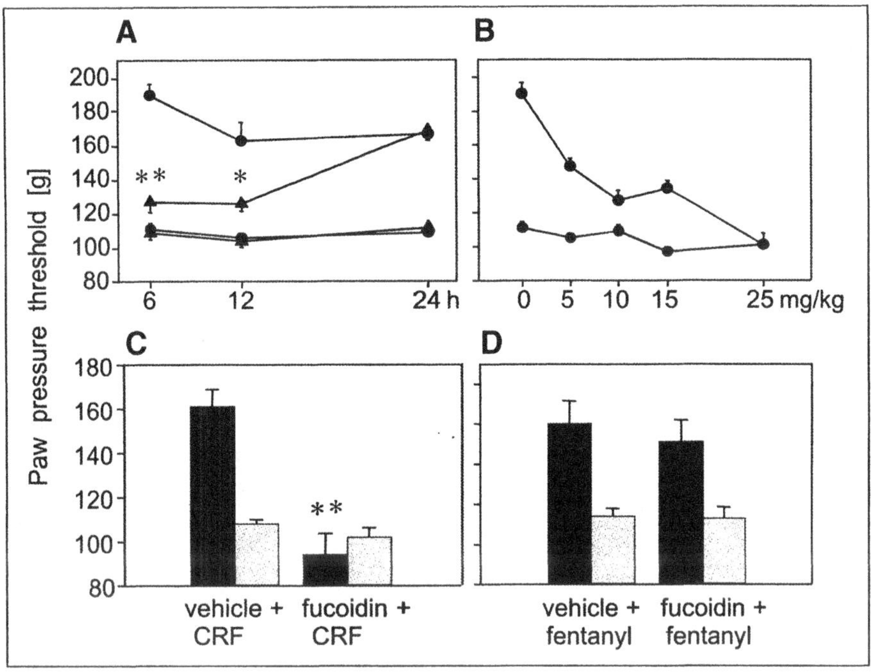

Fig. 6. The effect of fucoidin (L- and P-selectin blocker) on peripheral endogenous (A, B) and exogenous (C, D) opioid analgesia. (A) Time course of the effect of fucoidin (10 mg/kg) on cold water swim (CWS) stress-induced analgesia. Circles: control; triangles: fucoidin. Fucoidin significantly decreased analgesia at 6-12 hours after induction of inflammation. (B) Fucoidin dose-dependently decreased CWS-induced analgesia in inflamed but not in noninflamed paws ($p<0.001$ and $p>0.05$, regression ANOVA, respectively). (A, B) filled symbols: inflamed paws; open symbols: noninflamed paws. (C) Effects of fucoidin (10 mg/kg) on CRF (corticotropin releasing factor; 3 ng)- and (D) on fentanyl (3 μg)-induced analgesia. Black bars: inflamed paws; gray bars: noninflamed paws. Fucoidin significantly decreased CRF- but not fentanyl-induced analgesia. * $p<0.05$; ** $p=0.001$; Mann-Whitney test. Data are expressed as means ± SEM. Reprinted with permission from Machelska H et al. Nature Med 1998; 4:1425-1428.

Clinical Implications

Peripheral opioid analgesic actions are of clinical relevance. Opioid receptors are present on peripheral terminals of nerve fibers in human synovia[30] and these receptors are capable of mediating analgesia in humans.[31] Opioid peptides were found in human synovial lining cells and in immune cells such as lymphocytes, macrophages and mast cells. The prevailing peptides are β-endorphin and enkephalin, while only minor amounts of dynorphin are detectable.[30] These morphological findings are presented in details in the chapter by Mousa in this volume. The interaction of synovial opioids with peripheral opioid receptors was examined in patients undergoing knee surgery. Blocking intraarticular opioid receptors by the local administration of naloxone resulted in significantly increased postoperative pain.[32] Taken together, these findings suggest that in a stressful (e.g., postoperative) situation, opioids are tonically released from inflamed tissue and activate peripheral opioid receptors to attenuate clinical pain.

Importantly, these endogenous opioids do not interfere with exogenous morphine, i.e., intraarticular morphine is an equally potent analgesic in patients with and without opioid-producing inflammatory synovial cells.[30] This suggests that, in contrast to the rapid development of tolerance in the central nervous system, the immune cell-derived opioids do not readily produce cross-tolerance to morphine at peripheral opioid receptors.

These findings have to be kept in mind when the immune system becomes a target for the treatment of inflammatory diseases. The blockade of immune cell extravasation by antibodies against immunocytes or against adhesion molecules has been proposed as a novel anti-inflammatory strategy.[33] However, as shown above, anti-selectin treatment can cause severe impairment of opioid-mediated pain inhibition in inflammation.[23] Thus, it is important that future therapeutic strategies aimed at limiting the adhesion of harmful cells in inflammatory diseases do not interfere with the migration of opioid-containing cells promoting pain control.

Conclusions

In conclusion, opioid peptides released from immune cells under stress or by CRF can activate opioid receptors on sensory nerves to produce potent peripheral, clinically relevant analgesia. This analgesia lacks central side effects such as respiratory depression, sedation, dysphoria and dependence. It appears that the immune system uses mechanisms of cell migration not only to fight pathogens but also to control pain within injured tissue. Thus, pain may be exacerbated by measures inhibiting the immigration of opioid-producing cells or, conversely, analgesia may be conveyed by adhesive interactions that recruit those cells to injured tissue. These findings provide new insights into pain associated with a compromised immune system, as in AIDS or in cancer. The activation of opioid production and release from immune cells may be a novel approach to the development of peripherally acting analgesics. Since such drugs would be targeted towards events in peripheral injured tissue, these analgesics should lack unwanted central side effects typically associated with opioids.

Acknowledgments

Supported by grants from the Deutsche Forschungsgemeinschaft and the National Institutes of Health.

References

1. Machelska H, Binder W, Stein C. Opioid receptors in the periphery. In: Kalso E, McQuay H, Wisenfeld-Hallin Z, eds. Opioid sensitivity of chronic noncancer pain. Seattle: IASP Press, 1999:45-58.
2. Barber A, Gottschlich R. Opioid agonists and antagonists: an evaluation of their peripheral actions in inflammation. Med Res Rev 1992; 12:525-562.
3. Stein C. Peripheral mechanisms of opioid analgesia. Anesth Analg 1993; 76:182-191.
4. Stein C. The control of pain in peripheral tissue by opioids. N Engl J Med 1995; 332:1685-1690.
5. Stein C, Machelska H, Binder W et al. Peripheral opioid analgesia. Curr Opin Pharmacol 2001; 1:62-65.
6. Walker JS, Wilson JL, Binder W et al. The anti-inflammatory effects of opioids: their possible relevance to the pathophysiology and treatment of rheumatoid arthritis. Res Alerts Rheum Arthritis 1997; 1:291-299.
7. Höllt V. Opioid peptide processing and receptor selectivity. Annu Rev Pharmacol Toxicol 1986; 26:59-77.
8. Zadina JE, Hackler L, Ge L-J et al. A potent and selective endogenous agonist for the μ-opiate receptor. Nature 1997; 386:499-502.
9. Blalock JE. The syntax of immune-neuroendocrine communication. Immunol Today 1994; 15:504-511.
10. Weisinger G. The transcriptional regulation of the preproenkephalin gene. Biochem J 1995; 307:617-629.
11. Panerai AE, Sacerdote P. β-endorphin in the immune system: a role at last? Immunol Today 1997; 18:317-319.

12. Lyons PD, Blalock JE. Proopioimelanocortin gene expression and protein processing in rat mononuclear leukocytes. J Neuroimmunol 1997; 78:47-56.
13. Przewlocki R, Hassan AHS, Lason W et al. Gene expression and localization of opioid peptides in immune cells of inflamed tissue. Functional role in antinociception. Neurosci 1992; 48:491-500.
14. Cabot PJ, Carter L, Gaiddon C et al. Immune cell-derived β-endorphin: production, release and control of inflammatory pain in rats. J Clin Invest 1997; 100:142-148.
15. Mousa SA, Zhang Q, Sitte N et al. β–Endorphin containing memory-cells and μ-opioid receptors undergo transport into peripheral inflamed tissue. J Neuroimmunol 2001; 115:71-78.
16. Rittner HL, Brack A, Machelska H et al. Opioid peptide expressing leukocytes-identification, recruitment and simultaneously increasing inhibition of inflammatory pain. Anesthesiology 2001; 95:500-508.
17. Hassan AHS, Przewlocki R, Herz A et al. Dynorphin, a preferential ligand for kappa-opioid receptors, is present in nerve fibers and immune cells within inflamed tissue of the rat. Neurosci Lett 1992; 140:85-88.
18. Stein C, Schäfer M, Cabot PJ et al. Peripheral opioid analgesia. Pain Rev 1997; 4:171-185.
19. Springer TA. Traffic signals for lymphocyte recirculation and leukocyte emigration: the multistep paradigm. Cell 1994; 76:301-314.
20. Butcher EC, Picker LJ. Lymphocyte homing homeostasis. Science 1996; 272:60-66.
21. Petruzzelli L, Takami M, Humes D. Structure and function of cell adhesion molecules. Am J Med 1999; 106:467-476.
22. Mousa SA, Machelska H, Schäfer M et al. Co-expression of β-endorphin with adhesion molecules in a rat model of inflammatory pain. J Neuroimmunol 2000; 108:160-170.
23. Machelska H, Cabot PJ, Mousa SA et al. Pain control in inflammation governed by selectins. Nat Med 1998; 4:1425-1428.
24. Schäfer M, Mousa SA, Stein C. Corticotropin-releasing factor in antinociception and inflammation. Eur J Pharmacol 1997; 323:1-10.
25. Schäfer M, Carter L, Stein C. Interleukin-1 beta and corticotropin-releasing-factor inhibit pain by releasing opioids from immune cells in inflamed tissue. Proc Natl Acad Sci USA 1994; 91:4219-4223.
26. Stein C, Millan MJ, Herz A. Unilateral inflammation of the hindpaw in rarts as a model of prolonged noxious stimulation: alternations in behavior and nociceptive thresholds. Pharmacol Biochem Behav 1988; 31:445-451.
27. Stein C, Gramsch C, Herz A. Intrinsic mechanisms of antinociception in inflammation. Local opioid receptors and beta-endorphin. J Neurosci 1990; 10:1292-1298.
28. Stein C, Hassan AHS, Przewlocki R et al. Opioids from immunocytes interact with receptors on sensory nerves to inhibit nociception in inflammation. Proc Natl Acad Sci USA 1990; 7:5935-5939.
29. Schäfer M, Mousa SA, Zhang Q et al. Expression of corticotropin-releasing factor in inflamed tissue is required for intrinsic peripheral opioid analgesia. Proc Natl Acad Sci USA 1996; 93:6096-6100.
30. Stein C, Pflüger M, Yassouridis A et al. No tolerance to peripheral morphine analgesia in presence of opioid expression in inflamed synovia. J Clin Inves 1996; 98:793-799.
31. Stein C, Comisel K, Haimerl E et al. Analgesic effect of intraarticular morphine after arthroscopic knee surgery. N Engl J Med 1991; 325:1123-1126.
32. Stein C, Hassan AHS, Lehrberger K et al. Local analgesic effect of endogenous opioid peptides. Lancet 1993; 342:321-324.
33. Choy EHS, Panayi GS, Kingsley GH. Therapeutic monoclonal antibodies. J Rheumatol 1995; 34:707-715.

CHAPTER 8

Opioid Receptor Expression and Intracellular Signaling by Cells Involved in Host Defense and Immunity

Burt M. Sharp

Abstract

More than two decades ago, Joseph Wybran reported his original insights on the expression of different opioid receptor types by T-cells. This was based on the differential effects that morphine and methionine enkephalin exerted on human T-cell rosetting in the presence of sheep red blood cells. Since that time, numerous laboratories have shown that opiate alkyloids and opioid peptides have pleiotropic effects on immune function. In general, these compounds act as immunomodulators that modify the immune response to mitogens, antigens and antibodies that cross-link the T-cell receptor. In the past decade, it has become clear that cells involved in host defense and immunity express the various mRNAs encoding the same opioid receptors originally identified in neuronal tissues. Recently, indirect fluorescence and immunofluorescence have been utilized to demonstrate the regulated expression of both delta and kappa opioid receptors, predominantly on T-cells. In addition, immune cells express sites that show atypical opiate and opioid binding properties. In this review, we will distill the evidence for both classical and atypical opioid receptors and their effects on signaling within immune cells, focusing on the T-cell and emphasizing the δ opioid receptor.

Introduction

The direct immunomodulatory effects of opiate alkyloids and opioid peptides have been recognized for more than two decades. Thus, β-endorphin and synthetic opioid peptide agonists selective for the δ opioid receptor (DOR) are known to modulate thymic and splenic T cell proliferation, cytokine production and calcium mobilization.[1-4] Selective DOR agonists, such as [D-Ala_2-D-Leu_5]-enkephalin (DADLE) and deltorphin, were shown to inhibit the anti-CD3-ε-induced proliferation of highly purified murine splenic $CD4^+$ and $CD8^+$ T-cells.[2] In addition, interleukin (IL)-2 production was partially suppressed at higher concentrations (10^{-8} – 10^{-6} M) of DADLE and deltorphin. Although these studies provide pharmacological evidence for the presence of DOR on T-cells, the mechanism(s) whereby these compounds modulate the activation and function of cells involved in host defense and immunity has been obscured by the lack of direct evidence

Immune Mechanisms of Pain and Analgesia, edited by Halina Machelska and Christoph Stein.

for the expression of opioid receptors by these cells. Recent studies from several laboratories have largely resolved this dilemma. This review will focus on observations that have advanced our understanding of the expression and intracellular effects of opioid receptors on lymphocytes and other cells in the peripheral immune system.

The original observations showing that morphine and methionine enkephalin directly affect immune function were made by Joseph Wybran et al.[5] His 1979 report demonstrated that these compounds had opposite, yet direct, effects on the rosetting of human T-cells to sheep red blood cells. Based on our current understanding, these effects on T-cell rosetting indicate that opiates modulate the CD2 receptor expressed by T-cells. Based on these disparate effects of morphine and methionine enkephalin, Wybran inferred that human T-cells express different subtypes of opioid receptors that show preferential interactions with and confer differential, and sometimes opposite, cellular responses to these agonists.

Identification of Classical and Atypical Opioid Receptors on Immune Cells by Radioligand Binding

A number of laboratories have reported radioligand binding studies of opiates to immune cells. In many of these, the data sets were incomplete or atypical binding sites were found. Using [^{3}H]-naloxone, experiments with freshly obtained human peripheral blood lymphocytes suggested the existence of a morphine-sensitive site, since approximately 40% of the [^{3}H]-naloxone was displaced by morphine concentrations in the nM range.[6] Amongst the atypical sites, Roy and colleagues reported morphine binding on resting thymocytes, with a relatively low affinity for morphine (approximate Kd = 100 nM).[7,8] Similar sites on human peripheral blood macrophages were reported by Makman et al who described a low affinity, naloxone-insensitive binding for morphine, designated μ_3.[9] Lastly, high affinity, naloxone- and morphine-insensitive binding of β-endorphin was extensively characterized on both murine splenocytes in culture and U937 cells, a human promonocytic cell line.[10,11] Our group reported that binding was saturable and sensitive to cations, guanosine triphosphate (GTP)-γS and incubation with phorbol myristate acetate.[10-12] In contrast to neuronal opioid receptors, the β-endorphin binding site required the peptide's C-terminus (β-endorphin$_{28-31}$) and N-acetyl-β-endorphin was virtually equipotent to β-endorphin. Thus, both morphine and β-endorphin can bind to lymphoid cells through naloxone-insensitive sites that are distinctly different, atypical, and appear to modulate immune functions related to cellular proliferation.[7,13,14]

A limited number of radioligand binding studies have detected δ- and κ opioid ligand binding to lymphocytes and other cells involved in host defense. Using [^{3}H]-deltorphin, binding to a single high affinity site (Kd = 0.45 nM) was observed on human peripheral blood polymorphonuclear leukocytes.[15] Moreover, pretreatment with the irreversible ligand [D-Ala^2, Leu^5, Cys^6]-enkephalinamide (Dalce-NH2) differentially affected the binding of [^{3}H]-deltorphin vs. [^{3}H]-[D-Pen^2, D-Pen^5] enkephalin (DPDPE), suggesting that these cells express two types of δ-like opioid receptors. Other investigators have reported that macrophages and lymphocytes could be covalently cross-linked with δ selective ligands. Thus, labeling with cis-(+)-3-methylfentanylisothiocyanate (SUPERFIT) suggested the presence of δ-like opioid receptors on both B- and T-cell-enriched murine splenocytes.[16] Lymphoid cell lines have been used in binding studies with κ selective ligands such as [^{3}H]-U69,593. The murine R1.1 thymoma cell line showed a single high affinity site that bound [^{3}H]-U69,593 or [^{3}H]-naloxone, and competition studies were consistent with expression of a κ opioid receptor (KOR).[17] Data from additional experiments with cations or various guanylyl nucleotides were consistent with the presence of a neuronal KOR that coupled to a G protein.[18] Taken together, these studies indicate the difficulty of detecting and adequately characterizing opioid receptors on normal immune cells using reversible radioligand binding. In contrast, cell lines have been used to describe classical neuronal KORs on cells derived from a murine thymoma.

Identification of Opioid Receptor Transcripts in the Immune System

Transcripts for μ, δ- and κ opioid receptors have been detected on a variety of immune cells using primarily reverse transcription with polymerase chain reaction (RT-PCR). Initially, DOR mRNA was identified in simian peripheral blood mononuclear cells.[19] Then, DOR transcripts were found in human T-, B- and monocyte cell lines and in some murine lymphocytic cell lines.[20] Our group reported[21] that the sequence of the PCR transcript amplified from 90%-pure murine lymph node and splenic T-cells (Balb/c mice) was 98% identical with the murine DOR mRNA originally reported by Evans et al, in 1992.[22] Freshly obtained murine splenocytes expressed very low transcript levels, and this was increased by culturing splenocytes without mitogenic stimulation.[21,23] Using quantitative competitive RT-PCR amplification, we subsequently observed approximately 1 DOR transcript per T-cell in extracts prepared from fresh T-cell enriched populations (CD1 mice) that were 85% positive for Thy-1.[24] Thus, the expression of murine DOR mRNA in mature T-cells does not appear to be strain-dependent.

Transcripts for μ and κ opioid receptors (MOR, KOR, respectively) have been identified by several laboratories. MOR was found in rat peritoneal macrophages and in human and simian peripheral blood mononuclear cells;[25] in humans, monocytes, granulocytes and CD4+ T-cells were positive.[26,27] KOR transcripts were also detected in human and monkey peripheral blood lymphocytes.[26,27] In addition, immature thymic $CD4^-$, $CD8^-$ T-cells also express KOR mRNA.[28]

Regulation of DOR Transcript Expression

Amongst the three opioid receptor types, the regulation of DOR transcript expression has been studied most extensively. We observed that cell culture, in the absence of mitogens, increased transcript expression by 10- and 20-fold after 24 and 48 hours, respectively.[24] In addition, DOR transcript expression increased linearly with increasing cell density when splenocytes were in culture for 48 h.[23] To determine whether induction was dependent on the secretion of a soluble factor, supernatant transfer experiments were performed in which cells were cultured at relatively low density (1.2×10^6 cells/cm^2) in medium containing supernatant from cells that previously had been cultured at either low or high density (3.0×10^6 cells/cm^2). DOR expression was not enhanced by supernatants from cells cultured at high density.[23] Therefore, direct cell-cell interactions appear to mediate the enhanced expression of DOR mRNA by cultured splenocytes in the absence of mitogenic stimulation.

Stimulation by the T-cell mitogen, concanavalin A,[29] or cross-linking the T-cell receptor (TCR) with anti-CD3-ε[24] enhanced the expression of DOR mRNA in murine $CD4^+$ T-cells. Using quantitative competitive PCR, our laboratory reported that anti-CD3-ε doubled the number of DOR transcripts per T-cell in comparison to the effect of culture alone.[24] Thus, copy number increased from 10 (culture alone) to 22 and from 20 to 42 at

24 and 48 hours, respectively. In the presence of actinomycin D, anti-CD3-ε failed to alter the rate of transcript degradation (apparent half-life of 6 hours), suggesting that stability was unaffected.[24] Hence, increased transcription appears to account for the anti-CD3-ε-induced expression of DOR mRNA.

Phorbol myristate acetate (PMA) in the presence or absence of ionomycin affects DOR transcript expression in a manner opposite to that observed with anti-CD3-ε. We recently found that PMA alone or in combination with ionomycin inhibited the enhanced expression of T-cell DOR mRNA observed when splenocytes were cultured in the absence of mitogen.[24] This was evident when T-cells were separated after splenocytes had been in culture for 24 and 48 hours. PMA also inhibited the anti-CD3-ε-induced expression of DOR transcripts. Inspection of the mouse DOR gene promoter region revealed the presence of two sequences similar to the consensus TRE (phorbol ester-responsive element) and TRE-like elements. Although the function of these TRE domains has not been determined, studies have demonstrated that both of these elements mediate activation,[30,31] as well as inhibition,[32] of gene expression, depending on the cellular context. In a wide variety of cell types, phorbol esters have been shown to inhibit gene transcription. For example, PMA apparently inhibited transcription of the gene encoding the P2U-purinergic nucleotide receptor, a 7-transmembrane G-protein-coupled receptor expressed by HL-60 cells.[33] Since the activation of protein kinase c (PKC) apparently inhibits DOR gene expression, but anti-CD3-ε (an activator of PKC) has the opposite effect, it is possible that other TCR-dependent intracellular effectors deliver a dominant positive signal(s) resulting in enhanced DOR expression. The activation of PKC by PMA may shift this balance, perhaps reflecting that different isoforms of PKC are activated by PMA and/or the duration and magnitude of activation of PKC by PMA differs from anti-CD3-ε.

Identification of KOR and DOR by Indirect Fluorescence and Immunofluorescence Labeling

Using a high affinity κ-agonist conjugated to fluorescein (FITC-AA) and amplified with biotinylated anti-fluorescein IgG and extravidin-R-phycoerythrin, KOR were detected on murine thymocytes by fluorescence flow cytometry.[34] More than 50% of the thymocytes positive for both CD4 and CD8 were KOR positive,[35] and most of the thymocytes expressing KOR were of this phenotype. In contrast, murine splenocytes expressed KOR on less that 25% of freshly obtained CD4+ T-cells and 16% of B lymphocytes.[36] These findings suggest a decrease in KOR expression with T-cell maturation. In recent studies, Bidlack et al showed that mitogenic stimulation increased KOR expression by splenic T-cells in culture; this was especially evident on the $CD8^+$ subset.[37]

Recent studies from our laboratory have identified DOR on splenocytes by immunofluorescence, using epifluorescence microscopy with digital image analysis.[38] Balb/c Vβ8.1 mice received a single injection of the superantigen staphylococcal enterotoxin B (SEB; which activates the TCR) and spleens were obtained at various time intervals, thereafter. SEB induced both DOR mRNA and enhanced immunofluorescence. Approximately 50% of the total T-cell population expressed DOR immunofluorescence within 15 hours of SEB treatment, compared to less than 10% of control T-cells. DOR expression was elevated for 24 hours, then gradually declined toward control levels by 72 hours. On a single cell basis, relative DOR immunofluorescence increased approximately two-fold in a subpopulation of T-cells (26.8 ± 8.6 % of all T-cells) whose fluorescence intensity was greater than 2 standard deviations above the mean value of the corresponding control group.

These studies demonstrate that DOR is expressed by T-cells in vivo through a TCR-dependent mechanism.

In recent flow cytometric studies, we have observed that phytohemagglutinin (PHA) stimulated the expression of DOR immunoflurescence by human peripheral blood T-cells.[39] Approximately 50% of both the $CD4^+$ and $CD8^+$ T-cell subsets expressed DOR by 48 hours, and more than 90% of DOR was found on these cells. In additional experiments, the cell surface markers CD45RA and CD45RO were used to determine whether DOR is expressed by naïve or memory T-cells, respectively.[39] These studies were performed 48 hours after stimulation by PHA, a time interval when the fraction of lymphocytes expressing $CD45RA^+$ and $CD45RO^+$ was unaffected by PHA. PHA stimulated similar fractions of both CD45RA and CD45RO positive T-cells to express DOR. Other studies (our unpublished observations) have shown that anti-CD3-ε-induced the expression of DOR immunofluorescence by $CD4^+$ and $CD8^+$ T-cells obtained after splenocyte culture.

Opioid Receptor-Mediated Intracellular Signaling in the Immune System

In neuronal tissues, adenylyl cyclase is known to be inhibited by all three opioid receptor types. As it can be difficult to examine intracellular signaling with heterogenous populations of lymphoid cells, cell lines have been utilized by many investigators. The KOR expressed by the R1.1 thymoma cell line has been shown to inhibit basal and forskolin-stimulate cAMP production in a pertussis toxin-sensitive manner, consistent with coupling to G_i proteins.[40] Using a human T-cell line that stably expressed DOR (DOR-Ju1.1), we also found that DADLE reduced forskolin-stimulated cAMP production by 70%, with an IC_{50} of 10^{-11} M, and this was sensitive to pertussis toxin.[41] Finally, studies with human peripheral blood lymphocytes reported a biphasic effect of methionine enkephalin on intracellular cAMP concentrations: low concentrations (e.g., 10^{-12} M) elevated cAMP within 15 min, whereas nM concentrations reduced cAMP levels by 2hours.[42] Unfortunately, the effect of antagonists was not determined.

Several studies on the phosphorylation of the mitogen-activated protein kinases (MAPKs), extracellular-regulated kinases (ERKs) 1 and 2, have been reported in lymphoid cell lines. We and others reported that the DORs expressed by DOR-Ju1.1 cells rapidly stimulate MAPK phosphorylation in a ras-independent manner.[43,44] We also found that herbimycin A, a protein tyrosine kinase inhibitor, reduced the DADLE-induced phosphorylation of ERKs 1 and 2 by 70%.[43] Furthermore, in CEMx174 lymphocytic cells, morphine stimulated the expression of proteins involved in the MAPK cascade and increased both the expression and phosphorylation of ERK 1 and 2.[45] Thus, both MOR and DOR are coupled through, as yet undefined, pathways which activate ERK 1 and 2 in lymphoid cell lines.

The aforementioned findings differ from the effects of DOR (and MOR) on normal mature T-cells. Recently, DADLE was shown to inhibit anti-CD3-ε-induced phosphorylation of ERKs 1 and 2 in murine splenic T-cells.[38] Staphylococcal enterotoxin B was initially administered in vivo to stimulate the expression of DORs. Spleens were harvested 24 hours thereafter, and mixed splenocytes were preincubated with DADLE for 60 min and then stimulated with anti-CD3-ε for 5 min. In a concentration-dependent manner, we observed that DADLE (nM) reduced ERK phosphorylation by as much as 50%. The kinetics of anti-CD3-ε-induced ERK phosphorylation were unaffected by DADLE, and DADLE alone did not alter ERK phosphorylation.[38] Thus, DOR activation appears to inhibit anti-CD3-ε-induced ERK phosphorylation, rather than accelerating dephosphorylation of the ERKs. As noted, these observations differ from those made with DOR-transfected Jurkat cells (DOR-Ju1.1), yet they fit well with a previous report showing that DADLE inhibited anti-CD3-ε-driven IL-2 production and proliferation of highly purified murine $CD4^+$ and $CD8^+$ T-cells. These inhibitory actions of DORs all require preincubation with DOR agonists prior to TCR cross-linking.

Summary

After more than two decades of intensive study, there is clear evidence from multiple laboratories for the existence and regulated expression of opioid receptors by cells involved in host defense and immunity. At both the protein and mRNA levels, there is especially good evidence for DOR and KOR on T-cells and other immune cells in several species. In the case of KOR, expression is greater in lymphoid compartments (e.g., thymus) that contain immature cells. However, both DOR and KOR expression are enhanced in activated cells. In addition, T-cells show enhanced DOR expression as a function of increasing cell density, independent of activation through the TCR. Although both MOR and DOR agonists can stimulate ERK phosphorylation in lymphoid cells lines, the opposite occurs in normal T-cells. These agonists alone do not affect ERK phosphorylation; however, following preincubation, they do inhibit anti-CD3-ε-induced ERK phosphorylation in splenic T-cells. Based on these observations, along with the anti-proliferative effects of DOR agonists, DORs appear to be part of an inhibitory immunomodulatory system that can respond to both opioid neuro- and immunopeptides secreted locally by innervating neurons and immune cells, respectively.

Acknowledgement

Our research is supported by PHS DA-04196.

References

1. Linner KM, Quist HE, Sharp BM. Met-enkephalin-containing peptides encoded by proenkephalin A mRNA expressed in activated murine thymocytes inhibit thymocyte proliferation. J Immunol 1995; 154:5049-5060.
2. Shahabi NA, Sharp BM. Anti-proliferative effects of delta opioids on highly purified CD_4^+ and CD_8^+ murine T-cells. J Pharmacol Exp Ther 1995; 273:1105-1113.
3. Gilmore W, Weiner LP. Beta-endorphin enhances interleukin-2 (IL-2) production in murine lymphocytes. J Neuroimmunol 1988; 18:125-138.
4. Shahabi NA, Heagy W, Sharp BM. β-Endorphin enhances concanavalin-stimulated calcium mobilization by murine splenic T cells. Endocrinology 1996; 137:3386-3393.
5. Wybran J, Appelboom T, Famaey J-P, et al. Suggestive evidence for receptors for morphine and methionine-enkephalin on normal human blood T-lymphocytes. J Immunol 1979; 123:1068-1070.
6. Mehrishi JN, Mills IH. Opiate receptors on lymphocytes and platelets in man. Clin Immunol Imunopathol 1983; 27:240-249.
7. Roy S, Ge BL, Ramakrishan S, et al. [^{3}H]-morphine binding to thymocytes is enhanced by IL-1 stimulation. FEBS Lett 1991; 287:93-96.
8. Roy S, Ramakrishman S, Loh HH, et al. Chronic morphine treatment selectively suppresses macrophage colony formation in bone marrow. Eur J Pharmacol 1991; 195:359-363.
9. Makman MH, Dvorkin B, Stefano GB. Murine macrophage cell lines contain μ_3-opiate receptors. Eur J Pharmacol 1995; 273:5-6.
10. Shahabi NA, Linner KN, Sharp BM. Murine splenocytes express a naloxone-insensitive binding site for β-endorphin. Endocrinology 1990; 126:1442-1448.
11. Shahabi NA, Peterson PK, Sharp BM. β-Endorphin binding to naloxone-insensitive sites on a human mononuclear cell line (U937): Effect of cations and guanosine triphosphate. Endocrinology 1990; 126:3006-3015.

12. Shahabi NA, Sharp BM. Activation of protein kinase C rapidly down-regulates naloxone-resistant receptors for β-endorphin on U937 cells. J Pharmacol Exp Ther 1993; 264:276-281.
13. Shahabi NA, Burtness MZ, Sharp BM. N-acetyl-β-endorphin-(1-31) antagonizes the suppressive effect of β-endorphin-(1-31) on murine splenocyte proliferation via a naloxone-resistant receptor. Biochem Biophys Res Commun 1991; 175:936-942.
14. Roy S, Sedqi M, Ramakrishnan S et al. Differential effects of opioids on the proliferation of a macrophage cell line. Cell Immunol 1996; 169:271-277.
15. Stefano GB, Melchiorri P, Negri L et al. [D-Ala2] deltorphin I binding and pharmacological evidence for a special subtype of delta opioid receptor on human and invertebrate immune cells. Proc Natl Acad Sci USA 1992; 89:9316-9320.
16. Carr DJ, Kim CH, DeCosta B et al. Evidence for delta-class opioid receptor on cells of the immune system. Cell Immunol 1988; 116:44-51.
17. Bidlack JM, Saripalli LD, Lawrence DMP. κ-opioid binding sites on a murine lymphoma cell line. Eur J Pharmacol 1992; 227:257-265.
18. Lawrence DMP, Bidlack JM. Kappa opioid binding sites on the R1.1 murine lymphoma cell line: sensitivity to cations and guanine nucleotides. J Neuroimmunol 1992; 41:223-230.
19. Chuang LF, Chuang TK, Killam KF et al. Delta opioid receptor gene expression in lymphocytes. Biochem Biophys Res Commun 1994; 202:1291-1299.
20. Gaveriaux C, Peluso J, Simonin F et al. Identification of kappa- and delta-opioid receptor transcripts in immune cells. FEBS Lett 1995; 369:272-276.
21. Sharp BM, Shahabi NA, McKean D et al. Detection of basal levels and induction of delta opioid receptor mRNA in murine splenocytes. J Neuroimmunol 1997; 78:198-202.
22. Evans CJ, Keith DE,N□orrison H et al. Cloning of a delta opioid receptor by functional expression. Science 1992; 258:1952-1955.
23. Sharp BM, Li MD, Matta SG et al. Expression of delta opioid receptors and transcripts by splenic T cells. Ann N Y Acad Sci 2000; 917:764-770.
24. Li MD, McAllen K, Sharp BM. Regulation of delta opioid receptor expression by anti-CD3-ε, PMA, and ionomycin in murine splenocytes and T cells. J Leukocyte Biol 1999;65:707-714.
25. Sedqi M, Roy S, Ramakrishnan S et al. Complementary DNA cloning of a μ-opioid receptor from rat peritoneal macrophages. Biochem Biophys Res Commun 1995; 209:563-574.
26. Chuang LF, Chuang TK, Killam KF et al. Expression of kappa opioid receptors in human and monkey lymphocytes. Biochem Biophys Res Commun 1995; 209:1003-1010.
27. Chuang LF, Killam KF, Chuang RY. Induction and activation of mitogen-activated protein kinases of human lymphocytes as one of the signaling pathways of the immunomodulatory effects of morphine sulfate. J Biol Chem 1997; 272:26815-26817.
28. Belkowski SM, Zhu J, Liu-Chen L-Y et al. Detection of kappa-opioid receptor mRNA in immature T-cells. Adv Exp Med Biol 1995; 373:11-16.
29. Miller B. Delta opioid receptor expression is induced by concanavalin A in CD4$^+$ T cells. J Immunol 1996; 157:5324-5328.
30. Casey JL, Di JB, Rao KK et al. Deletional analysis of the promoter region of the human transferrin receptor gene. Nucleic Acids Res 1988; 16:629-646.
31. Ouyang Q, Bommakanti M, Miskimins WK. A mitogen responsive promoter region that is synergistically activated through multiple signalling pathways. Mol Cell Biol 1993; 13:1796-1804.
32. Lok C, Chan KF, Loh TT. Transcriptional regulation of transferrin receptor expression during phorbol-ester-induced HL-60 cell differentiation. Evidence for a negative regulatory role of the phorbol-ester-responsive element-like sequence. Eur J Biochem 1996; 236:614-619.
33. Martin KA, Kertsey SB, Dubyyak GR. Down-regulation of P2U-purinergic nucleotide receptor mesenger RNA expression during in vitro differentiation of human myeloid leukocytes by phorbol esters or inflammatory activators. Mol Pharmacol 1997; 51:97-108.
34. Lawrence DMP, El-Hamouly W, Archer S et al. Identification of κ-opioid receptors in the immune system by indirect immunofluorescence. Proc Natl Acad Sci USA 1995; 92:1062-1066.
35. Ignatowski TA, Bidlack JM. Detection of kappa opioid receptors on mouse thymocyte phenotypic subpopulations as assessed by flow cytometry. J Pharmacol Exp Ther 1998; 284:298-306.
36. Ignatowski TA, Bidlack JM. Differential kappa-opioid receptor expression on mouse lymphocytes at varying stages of maturation and on mouse macrophages after selective elicitation. J Pharmacol Exp Ther 1999; 290:863-870.

37. Bidlack JM, Abraham MK. Mitogen-induced activation of mouse T cells increases kappa opioid receptor expression. In: Friedman H, ed. Neuroimmune Circuits, Drugs of Abuse and Infectious Disease. New York: Pienum Press, In press.
38. Shahabi NA, McAllen K, Matta SG et al. Expression of delta opioid receptors by splenocytes from SEB-treated mice and effects on phosphorylation of MAP kinase. Cellular Immunology 2000; 205:84-93.
39. Sharp BM, McAllen K, Gekker G et al. Immunofluorescence detection of delta opioid receptors (DOR) on human peripheral blood CD4⁺ T cells and DOR-dependent suppression of HIV-1 expression. J Immunol 2001; 167:1097-1102
40. Lawrence DMP, Bidlack JM. The kappa opioid receptor expressed on the mouse R1.1 thymoma cell line is coupled to adenylyl cyclase through a pertussis toxin-sensitive guanine nucleotide-binding regulatory protein. J Pharmacol Exp Ther 1993; 266:1678-1683.
41. Sharp BM, Shahabi NA, Heagy W et al. Dual signal transduction through delta opioid receptors in a transfected human T-cell line. Proc Natl Acad Sci USA 1996; 93:8294-8299.
42. Martin-Kleiner I, Osmak M, Gabrilovac J. Regulation of NK cell activity and the level of the intracellular cAMP in human peripheral blood lymphocytes by met-enkephalin. Exp Med (Berl.) 1991; 192:145-150.
43. Shahabi NA, Daaka Y, McAllen K et al. Delta opioid receptors expressed by stably transfected Jurkat cells signal through the map kinase pathway in a ras-independent manner. J Neuroimmunol 1999; 94:48-57.
44. Hedin KE, Bell MP, Huntoon CJ et al. Gi proteins use a novel βγ-and ras-independent pathway to activate extracellular signal-regulated kinase and mobilize AP-1 transcription factors in jurkat T l lymphocytes. J Biol Chem 1999; 274:19992-20001.
45. Chuang TK, Killam KF, Chuang LF et al. Mu opioid receptor gene expression in immune cells. Biochem Biophys Res Commun 1995; 216:922-930.

CHAPTER 9

Experimental Evidence for Immunomodulatory Effects of Opioids

Paola Sacerdote, Elena Limiroli and Leda Gaspani

Introduction

In recent years the experimental and clinical research has made it clear that the immune system does not stand alone, but it is profoundly affected by other organ systems, especially the central nervous and neuroendocrine systems. It is also increasingly clear that the immune system can in turn affect the functioning of these systems as well.[1]

The research on opioid peptides plays a fundamental role in the development of this awareness. Among the first experimental evidences suggesting the existence of a bidirectional link between brain and immune system we must include the observation by Wybran et al.[2] who in 1979 showed that morphine, a drug considered to alter only neuronal functions, could affect also the responses of human B lymphocytes. This report stimulated the research in the field of Neuroimmunomodulation, leading to the identification of several neuropeptide receptors on immune cells, and to the observations that many, if not all, neuropeptides, hormones and neurotransmitters can affect the immune responses.

However, although 21 years have passed and a wide literature is now available, the real significance of the role of opioids in the modulation of the immune system has not yet been ascertained.

Morphine and Endogenous Opioids

The problem of the significance of opioid-induced modulation of the immune system is complex and multifaceted. Two main conceptual problems arise: with the term opioid or opiate we indicate both endogenous opioid peptides, namely β-endorphin, met-enkephalin and dynorphin, and opiate analgesic drugs, such as morphine. It is evident that it is very different to investigate the existence of a physiological role for opioid peptides in the complex modulatory network operating on the immune system, or to evaluate the impact of a pharmacological morphine treatment. Moreover, due to the high affinity of morphine for the opioid receptors and to its higher stability in comparison with peptides, morphine has frequently been employed in an interchangeable way as a tool for the exploration of the immune effects of the opioids. The large majority of the studies with the opioid peptides have been conducted in vitro, often evaluating the effect on a

Immune Mechanisms of Pain and Analgesia, edited by Halina Machelska and Christoph Stein.

single cell population or cell line/clone. In these conditions the results have often been contradictory, depending on the doses and timing of opioid addition to cell cultures. For these series of reasons this overview will be focussed mainly on "ex vivo" experiments, where the administration of opiates or the manipulation of the opioid system was achieved in vivo.

Immunomodulation by Morphine

In vivo Studies (Table 1)

In vivo administration of morphine to rodents induces a decrease of multiple immune parameters, affecting almost all different cell types of the immune system. Natural killer cells (NK) have been shown to be very sensible to morphine modulation in vivo. Injections of morphine have been found to lead to depressed NK in rats, mice and monkeys.[3-11] There is considerable evidence that morphine given in vivo modulates T cell functions. Independently of the stimulus utilized (policlonal mitogens, CD3 activation, antigen specific challenge) T lymphocyte proliferation is decreased by both acute and chronic morphine administration.[3,4,10-16] In contrast, the effects of morphine on B lymphocytes are less evident.[11,17,18] Morphine at high concentrations has been shown to reduce the mitogenic response of splenic B cells[4,16] and an inhibited induction of antibody forming cells to sheep red blood cells was observed in different species.[2,18] However, the formation of an antibody response to sheep red blood cells requires interaction of macrophages, T cells and B cells. It was for example shown by Weber et al that morphine inhibited antibody responses to a T-dependent, but not to a T-independent antigen[19] suggesting that morphine did not affect B cell-function directly.

Clear inhibitory effects of in vivo morphine on monocyte/macrophages functions have been consistently described. Chronic morphine treatment inhibited proliferation and differentiation of Macrophage Colony Stimulating Factor (MCSF)-dependent macrophage progenitor cells in the bone marrow.[20] Phagocytosis of different pathogens was also impaired by morphine.[11,21-23] The impairment was evident with peritoneal, alveolar or splenic macrophages, indicating a general down-regulation of innate immunity.

The effects of in vivo morphine on T cell and macrophage cytokines are not well clarified and relatively little explored. Morphine seems to decrease interleukin (IL)-2 and interferon (IFN)-γ production by T lymphocytes.[5,10,24] Moreover, an increase of IL-4 production has been reported after morphine administration in some cases[25] and a decreased IL-4 in other experiments.[26] Also contradictory are the results reported of morphine modulation of IL-12 and IL-10, mostly depending on the time of cytokine measurement after administration of morphine and of the cell population analyzed. We observed a decrease of IL-10 and IL-12 production by murine peritoneal macrophages collected one hour after morphine administration (unpublished data), while an increase of the last cytokine produced by splenocyte cultures was reported 24 hours after morphine.[27]

In most of the results obtained with morphine a classical μ type opiate receptor is involved, since the effects of morphine can be blocked by the antagonist naloxone.[18] Moreover, the immunosuppressive effects of morphine disappear in mice knock-out (KO) for the μ opioid receptor.[28] Generally, however, the morphine doses needed in order to exert a significant immunosuppression are higher than the ones necessary to induce antinociception (analgesia). While in fact in the BALB-C mouse a relevant analgesic effect is observed at doses starting from 2.5 mg/kg, the doses generally employed in the experiments of immunosuppression start from 10 mg/kg.[10] Convincing evidence of the immunosuppressive effect of morphine comes from several experiments associating opioids to

Table 1. Morphine effects on immune cells

Cell Type	Effect	In vivo Studies	In vitro Studies
T-lymphocytes	Mitogen-induced proliferation	↓	↓–
	Interferon-γ production	↓	
	Interleukin-2 production	↓	
	Interleukin-4 production	↓↑	
	Interleukin-10 production	↑	
	T rosette formation		↓
	Apoptotic death		↑
B-lymphocytes	Mitogenic response	↓	
	Induction of Ab forming cells to sheep erythrocytes	↓	
Natural killer lymphocytes	Cytotoxicity	↓	–
Monocytes/macrophages	Proliferation of macrophage progenitor cells	↓	
	Differentiation of macrophage progenitor cells	↓	
	Phagocytosis	↓	
	Interleukin-12 production	↓	
	Interleukin-10 production	↓	
	Host resistance to infection disease agents	↓	

Key: ↓ inhibition; ↑ stimulation; – no effect, in vivo, in vivo administration of morphine

microbial pathogen infections.[29] It has in fact consistently been reported that acute as well chronic administration of morphine to rodent modulates host resistance to bacteria and fungal pathogens. A single dose or repeated doses of morphine have been shown to enhance the lethality of Toxoplasma,[30] Klebsiella and Candida.[21] Moreover, morphine can alter pathogenesis also in virus infection models, including murine Leukemia, Moloney sarcoma, Herpes simplex and Friend viruses.[31] It appears from all these studies that opiates act to alter host-defenses against a variety of infection disease agents. However, the outcomes of these infections are influenced by the timing of morphine administration, the state of opiate dependence, the animals species and the dose and route of infection.[29] These observations are indeed particularly important in the consideration of association between AIDS epidemic and opiate abuse in the human.

In vitro Studies (Table 1)

Although the first and pioneering study on the effects of morphine on the immune system described that morphine added in vitro to human peripheral blood lymphocytes inhibited T rosette formation,[2] the subsequent studies in the experimental animal did not consistently show a direct effect on different cell populations, since in general either a decrease or no modification of the immune parameters studied were observed.

While NK activity did not seem to be affected even by high morphine concentrations in vitro[32,33] reduction as well as no effect on lymphocyte proliferation was reported, depending on the mitogen and the source of T cells.[5,34,35] However, an interesting paper from Yin et al[36] demonstrated that morphine added in vitro can facilitate the triggering of apoptotic death of T lymphocytes by modulating the Fas-FasL system. These effects were mediated by opioid receptors present on the immune cells. The discrepancy between the in vitro and in vivo effects of morphine has prompted to the search of the mechanism involved in the morphine-induced effects.

Mechanisms of Morphine-Induced Immunosuppression

A good agreement has been reached that opioids induce immunosuppression either interacting directly with opioid receptors on immune cells or on receptors within the central nervous system.[37-41] It has been demonstrated that immune cells express μ, δ and κ receptors functionally coupled to signal transduction mechanisms.[42] Moreover activation of central opioid receptors can regulate the peripheral immune system throughout the stimulation of the hypothalamus-pituitary-adrenal axis (HPA)[33] and the sympathetic nervous system.[15] The activation of the HPA elicits the production of adrenocorticotropin from the pituitary that in turn elicits the release of glucocorticoids that could eventually suppress the immune system[33,43] Both primary and secondary lymphoid organs (such as the spleen) are innervated by the sympathetic nervous system.[44] Activation of this system by opioids elicits the release in these organs of catecholamines that have been demonstrated to suppress lymphocyte, NK cell and macrophage functions.[45]

Opioid Peptides

While the studies examining morphine-induced immunomodulation are quite consistent in claiming for immunosuppressive properties, the evidences collected on the role of the endogenous opioid peptides in immune responses are more complex and not definitive.

In vitro Studies (Table 2)

The μ and δ agonists, β-endorphin and met-enkephalin exert surprising effects in vitro on NK activity and T lymphocyte proliferation. The majority of the studies conducted, in fact, indicates that these peptides mostly increase these two parameters.[32,46-52] The stimulation was observed with splenocytes,[51] lymphonode cells[53,54] and peripheral blood lymphocytes in different animal species.[55,56] On the other hand, the entity and the duration of the stimulatory properties were often dependent on the experimental scheme, such as the timing of the peptide addition to cell cultures, i.e., before or after mitogen, and on the concentrations used. Sometimes biphasic responses were observed with low or high concentrations.[57] Macrophage functions are indeed either decreased or stimulated by in vitro administration of the opioid peptides. Phagocytosis is consistently impaired by μ, δ and κ ligands[58-60] while superoxide release was shown to be increased by β-endorphin,[61,62] dynorphin[61,62] and met-enkephalin.[63,64] An interesting hypothesis suggested is that opioid peptides and β-endorphin in particular could be fine regulators of immune function normalizing the immune responses, enhancing them when suppressed and reducing them when stimulated.[65]

In vivo Studies (Table 2)

As already reported for morphine, the data obtained with the in vivo administration of opioid peptides seem more consistent. When the opioid peptide β-endorphin was administered in vivo to rodents, either centrally in the brain or peripherally, a decrease of cellular immune functions, such as mitogen-induced lymphoproliferation and NK activity,

Table 2. Opioid peptide effects on immune cells

Cell Type	Effect	In vivo Studies	In vitro Studies
T-lymphocytes	Mitogen-induced proliferation	↓	↑
	Interferon-γ production	↓	
	Interleukin-2 production	↓	
	Interleukin-4 production	↑	
	Nitric oxide production	↑	
Natural killer lymphocytes	Cytotoxicity	↓	↑
Monocytes/macrophages	Phagocytosis		↓
	Respiratory burst activity		↑
	Survival of skin graft	improvement	
	Experimental Autoimmune Encephalitis	improvement	
	Inflammation	improvement	

Key: ↓ inhibition; ↑ stimulation; – no effect; in vivo, in vivo administration of opioid peptide

was observed.[66,67] On the contrary, the block of β-endorphin activity, achieved either with the administration of the opioid antagonists naloxone and naltrexone or of immunoglobulins that neutralize the activity of β-endorphin, induced an increase of NK activity and lymphoproliferation within minutes.[66,68] The in vivo administration of the opioid antagonists also was shown to affect in a negative way nitric oxide (NO) production by splenocytes, and this effect was demonstrated to be mediated mainly by central opioid receptors.[69] These observations, indicating that the removal of the opioid tone by way of an antagonist or an antibody can affect some immune responses, suggest that an endogenous opioidergic tone on some immune functions exists. The existence of this control has been further demonstrated in a model of inflammation in the rat. Indeed, administration of either β-endorphin or morphine affected the development of carrageenan-induced inflammation of rat paw. These opioid agonists in fact significantly reduced edema and cell infiltration in the inflamed paw, while naloxone or the administration of a neutralizing antibody against β-endorphin worsened the degree of edema.[79]

In the last years mice lacking μ, δ and κ opioid receptors have been generated and could provide tools to study the opioid impact on the immune system. Unfortunately, however, the data collected until now with these animals do not allow drawing final conclusions, since contradictory results have been published. Gaveriaux-Ruff et al[28] using a strain of μ-receptor KO mice did not show any difference between wild type and mutant mice with regard to NK activity or lymphocyte number. In contrast Tian et. al.[70] in a different μ-opioid receptor KO mice, reported the existence of an increased proliferation of granulocyte, macrophage, erytroid and multipotential progenitor cells in both bone marrow and spleen, indicating a link between hematopoiesis and the opioid system, suggesting that endogenous opioids exert a negative regulatory influence on immune functions. Many reasons such as the different background, or the fact that the absence of one receptor type could be compensated by the other remaining opioid receptors could explain the discrepancy in the results. More studies with double or triple opioid receptor KO need to be performed.[71]

Modulation of Th1/Th2 Responses

In vivo, opioid peptides have been involved in the modulation of the Th1/Th2 balance. It is well known that T helper cells are functionally polarized in two different subsets.[72] The Th1 and Th2 cells produce different patterns of cytokines: Th1 cells produce mainly IL-2, IFN-γ and lymphotoxins whereas Th2 cells produce IL-4, IL-5, IL-6, IL-10 and IL-13 (Fig. 1). Th1 cells are mostly involved in cells mediated reactions, while the Th2 cytokines are commonly found in association with strong antibody and allergic responses. Moreover, the characteristic cytokine products of Th1 and Th2 cells are inhibitory for the differentiation and effector function of the opposite subset.[73]

The administration of the opioid antagonist naloxone profoundly affected the splenocyte production of the Th1 cytokines IFN-γ and IL-2 and of the Th2 cytokine IL-4 in different strains of mice immunized with the protein antigen Keyhole Limpet Hemocyanin (KLH).[74] In fact, acute as well chronic naloxone treatment decreased IL-4 production. On the contrary, IL-2 and IFN-γ levels were increased after naloxone administration. The administration of naloxone seems therefore to stimulate Th1 cytokines, while decreasing Th2 cytokines. Given the fact that naloxone is an almost pure antagonist at the μ-opioid receptor, devoid of any intrinsic activity, the effects of the drug are likely to be due to the removal of a regulatory tone exerted by endogenous opioid peptides (Fig. 1).[75]

Since an imbalance of Th1/Th2 cytokines is often at the basis of immune diseases, the effect of naloxone and/or β-endorphin on immune responses can be relevant.[73] Therefore in the next paragraphs we report a few examples of how both opioids and the antagonism of opioid activities can affect the onset and the development of pathologies due to an altered immune functionality.

The Th1/Th2 paradigm has been involved in immune responses to organ transplantation.[76] Under certain experimental conditions in fact, graft rejection has been associated with the presence of Th1 cytokines, while Th2 cytokines have been linked to graft survival. We investigated the effect of the opioid receptor antagonist naloxone on the onset of allograft skin rejection in mice and on the production of IL-2, IFN-γ and IL-4 during the development of the rejection. The continuous administration of β-endorphin significantly prolonged the time of rejection while naloxone significantly shortened it.[77] This effect could be due to the effects of β-endorphin or naloxone on the Th1/Th2 balance. In fact, naloxone treatment increased Th1 cytokine production by splenocytes of transplanted mice evaluated at the moment of rejection.[77]

A possible involvement of opioids in an experimental model of autoimmune disease has also been suggested. Rat experimental autoimmune encephalitis (EAE) is commonly considered a model for human multiple sclerosis. The development of EAE is associated with Th1 responses, while remission of symptoms is linked to the appearance of a Th2 protective immune response. Consistently with the role of opioids in the modulation of Th1/Th2 responses, the administration of the opioid receptor antagonist naloxone, by increasing Th1 responses and blocking Th2 cytokines, worsened the clinical signs of EAE and increased mortality.[78]

Involvement of Opioids in Stress-Induced Immunosuppression

Recent research has provided convincing evidence that physical and psychological stress can affect the immune function both in the human and in the experimental animal.[80] During the period of stress (perceived as a threat to the homeostasis of the organism) the brain releases a number of chemical mediators, including opioid peptides, which

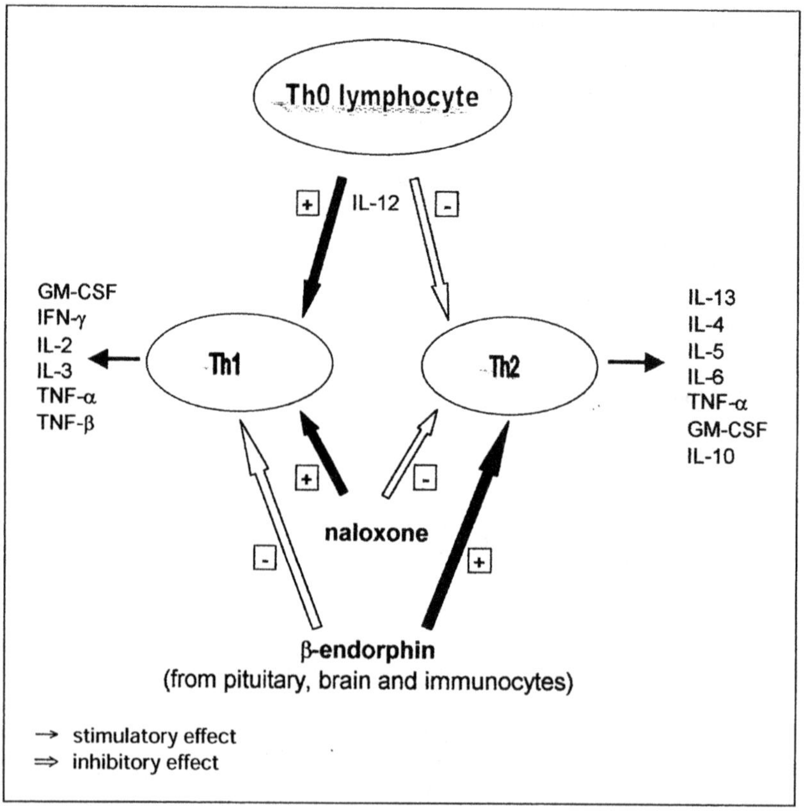

Fig. 1. Modulatory effects of β-endorphin and the opioid antagonist naloxone on T helper-1 and T helper-2 cytokine production. Abbreviations: Th, T-helper; IL, interleukin; GM-CSF, granulocyte macrophage colony stimulating factor; IFN-γ, interferon-γ; TNF, tumor necrosis factor.

affect many body systems comprising the immune system. In experiments dating back to 1984 Shavit et al[81] showed that the experimental stress model of intermittent footshock in the rat induced immunosuppression. Interestingly, only stress modalities, that were able to induce the release of opioid peptides, β-endorphin in particular, were associated with immunosuppression. Moreover, these stress-induced immune effects were blocked by the administration of the opioid antagonist naloxone. More recently, our group demonstrated that the administration of an antibody that neutralizes β-endorphin could block the immunosuppression induced by different stress models in the rat.[82,83] We have also shown that the β-endorphin mediating stress-induced immunosuppression is not only the one released by the pituitary, but also the peptide produced by immune cells themselves under stressful conditions. In this way the peptide can regulate the immune response by an autocrine/paracrine pattern.[83]

Conclusions

Considering the large amount of data collected in the last years on the immunomodulatory properties of opiates, it is clear that this activity must be added to the many pharmacological and physiological effects exerted on the central and peripheral nervous system, such as on nociceptive pathways and behavior, and on the neuroendocrine and gastroenteric systems. While the pharmacological alterations of immune responses induced by exogenous opiates, i.e., morphine and congeners, point to a final immunosuppression, the physiological role of endogenous opioids has not been completely defined. It becomes in fact questionable to claim for a unique immunosuppressive or immunostimulatory role for opioid peptides like β-endorphin. It can be suggested that the opioids might exert an inhibitory control of some cell populations (Th1, macrophages) probably throughout the stimulation of other cell types such as Th2 cells. Depending on the immune function evaluated (e.g., cellular vs. humoral), on the pre-existing Th1/Th2 balance (e.g., after previous exposure of animals to different pathogens or to genetic predisposition), on the stage of activation of immune cells, inhibition or stimulation of classical laboratory immune parameters can be achieved. If, however, we take into consideration the experiments with in vivo modulation of immune responses, it seems reasonable to suggest that the final outcome of the different inhibitory/stimulatory inputs exerted by opioids is that of a control aimed to prevent overshooting and excessive activation of the immune system.

References

1. Reichlin S. Neuroendocrine-immune interactions. N Engl J Med 1993; 329:1246-53.
2. Wybran J, Appelboom T, Farmaey JP. Suggestive evidence for receptors for morphine and methionine-enkephalin on normal human blood T lymphocytes. J Immunol 1979; 123:1068-1070.
3. Lysle DT, Coussons ME, Watts VJ et al. Morphine-induced alterations of immune status: dose dependency, compartment specificity and antagonism by naltrexone. J Pharmacol Exp Ther 1993; 265:1071-8.
4. Lysle DT, Hoffman KE, Dykstra LA. Evidence for the involvement of the caudal region of the periaqueductal gray in a subset of morphine-induced alterations of immune status. J Pharmacol Exp Ther 1996; 277:1533-40.
5. Fecho K, Maslonek KA, Dykstra LA et al. Assessment of the involvement of central nervous system and peripheral opioid receptors in the immunomodulatory effects of acute morphine treatment in rats. J Pharmacol Exp Ther 1996; 276:626-36.
6. Carr DJ, France CP. Immune alterations in morphine-treated Rhesus monkeys. J Pharmacol Exp Ther 1993; 267:9-15.
7. Carr DJ, Gebhardt BM, Paul D. α-adrenergic and IL-2 opioid receptors are involved in morphine-induced suppression of splenocyte natural killer activity. J Pharmacol Exp Ther 1993; 264:1179-86.
8. Carr DJ, Mayo S, Gebhardt B et al. Central α-adrenergic involvement in morphine-mediated suppression of splenic natural killer activity. J Neuroimmunol 1994; 53:53-63.
9. Carr DJ, Gerak LR, France CP. Naltrexone antagonizes the analgesic and immunosuppressive effects of morphine in mice. J Pharmacol Exp Ther 1994; 269:693-8.
10. Sacerdote P, Manfredi B, Mantegazza P et al. Antinociceptive and immunosuppressive effects of opiate drugs: a structure-related activity study. Br J Pharmacol 1997; 121:834-40.
11. Eisenstein TK, Hillburger ME. Opioid modulation of immune responses: effects on phagocyte and lymphoyd cell populations. J Neuroimmunol 1998; 83:36-44.
12. Bayer BM, Daussin S, Hernandez M et al. Morphine inhibition of lymphocyte activity is mediated by an opioid dependent mechanism. Neuropharmacol 1990; 29:369-74.
13. Chuang LF, Killam KF, Chuang RY. Opioid dependency and T-helper cell functions in rhesus monkey. In Vivo 1993; 7:159-166.
14. Flores LR, Hernandez MC, Bayer BM. Acute immunosuppressive effects of morphine: lack of involvement of pituitary and adrenal factors. J Pharmacol Exp Ther 1994; 268:1129-34.
15. Flores LR, Dretchen KL, Bayer BM. Potential role of the autonomic nervous system in the immunosuppressive effects of the acute morphine administration. Eur J Pharmacol 1996; 318:437-46.
16. Bryant HU, Roudebush RE. Suppressive effects of morphine pellet implants on in vivo parameters of immune function. J Pharmacol Exp Ther 1990; 255:410-414.
17. Bussiere JL, Adler MW, Rogers TJ et al. Effects of in vivo morphine treatment on antibody responses in C57 BL/6 bgJ/bgJ mice. Life Sci 1993; PL 52:43-48.

18. Flores RG, Weber RJ. Opioids, Opioids receptors and the Immune system. In: Plotnikoff, Faith, Murgo, Good, eds. Cytokines, Stress and Immunity. CRC Press, 1999:281-314.
19. Weber RJ, Ikejiri B, Rice KC et al. Opiate receptor mediated regulation of the immune response in vivo. NIDA Res Mono 1987; 76:341-348.
20. Roy S, Ramakrishnan S, Loh HH et al. Chronic morphine treatment selectively suppresses macrophage colony formation in bone marrow. Eur J Pharmacol 1991; 195:359-63.
21. Tubaro E, Borelli G, Croce C et al. Effect of morphine on resistance to infection. J Infect Dis 1983; 148:656-66.
22. Tubaro E, Santiangeli C, Belogi L et al. Methadone vs morphine: comparison of their effect on phagocytic functions. Int J Immunopharmacol 1987; 9:79-88.
23. Rojavin M, Szabo I, Bussiere JL et al. Morphine treatment in vitro or in vivo decreases phagocytic functions of murine macrophages. Life Sci 1993; 53:997-1006.
24. Scott M, Carr DJ. Morphine suppresses the alloantigen-driven CTL response in a dose-dependent and naltrexone reversible manner. J Pharmacol Exp Ther 1996; 278:980-88.
25. Roy S, Charboneau RG, Barke RA. Morphine synergizes with lipopolysaccharide in a chronic endotoxemia model. J Neuroimmunol 1999; 95:107-114.
26. Casalinuovo IA, Graziano R, Di Francesco P. Cytokine secretion by murine spleen cells after inactivated Candida albicans immunization. Effect of cocaine and morphine treatment. Immunopharmacol Immunotoxicol 2000; 22:35-48.
27. Pacifici R, Di Carlo S, Bacosi A et al. Pharmacokinetics and cytokine production in heroin and morphine-treated mice. Int J Immunophar 2000; 22:603-614.
28. Gaveriaux-Ruff C, Matthes HWD, Peluso J et al. Abolition of morphine-immunosuppression in mice lacking the mu opioid receptor gene. Proc Natl Acad Sci USA 1998; 95:6326-6328.
29. Risdahl JM, Khanna KV, Peterson PK et al. Opiates and infection. J Neuroimmunol 1998; 83:4-18.
30. Chao CC, Sharp BM, Pomeroy C et al. Lethality of morphine in mice infected with Toxoplasma gondii. J Pharmacol Exp Ther 1990; 252:605-609.
31. Starec M, Rouveix B, Sinet M et al. Immune status and survival of opiate and cocaine treated mice infected with Friend virus. J Pharmacol Exp Ther 1991; 259:745-750.
32. Mathews PM, Froelich CJ, Sibbit WL et al. Enhancement of natural cytotoxicity by beta-endorphin. J Immunol 1983; 130:1658-1662.
33. Freier DO, Fucks BA. A mechanism of action for morphine induced immunosuppression: corticosterone mediates morphine induced suppression of NK cell activity. J Pharmacol Exp Ther 1993; 270:1127-1133.
34. Palm S, Mignar T, Kuhn K et al. Phytohemagglutinin-dependent T -cell proliferation is not impaired by morphine. Methods Find Exp Clin Pharmacol 1996; 18:159-165.
35. Bayer BM, Gastonguay MR, Hernandez MC. Distinction between the in vitro and in vivo inhibitory effects of morphine on lymphocyte proliferation based on agonist sensitivity and naltrexone reversibility. Immunopharmacol 1992; 23:117-124.
36. Yin D, Mufson RA, Wang R et al. Fas-mediated cell death promoted by opioids. Nature 1999; 397:218.
37. Weber RJ, Pert A. The periaqueductal gray matter mediates opiate-induced immunosuppression. Science 1989; 245:188-190.
38. Guan L, Towsend R, Eisenstein TK et al. The cellular basis for opioid-induced immunosuppression. Adv Exp Med Biol 1995; 373:57-64.
39. Carr DJ, Rogers TJ, Weber RJ. The relevance of opioid receptors on immunocompetence and immune homeostasis. Proc Soc Exp Biol Med 1996; 213:248-57.
40. Shavit Y, Depaulis A, Martin FC et al. Involvement of brain opiate receptors in the immune-suppressive effect of morphine. Proc Natl Acad Sci USA 1986; 83:7114-7.
41. Flores LR, Dretchen KL, Bayer BM. Potential role of the autonomic nervous system in the immunosuppressive effects of acute morphine administration. Eur J Pharmacol 1996; 318:437-46.
42. Sharp BM, Roy S, Bidlack JM. Evidence for opioid receptors on cells involved in host defense and the immune system. J Neuroimmunol 1998; 83:45-56.
43. Bryant HU, Bernton EW, Kenner JR et al. Role of adrenal cortical activation in the immunosuppressive effects of chronic morphine treatment. Endocrinology 1991; 128:3253-58.
44. Hall DM, Suo J, Weber RJ. Opioid mediated effects on the immune system: sympathetic nervous system involvement. J Neuroimmunol 1998; 83:29-35.
45. Mellon RD, Bayer BM. Evidence for central opioid receptors in the immunomodulatory effects of morphine: review of potential mechanisms of action. J Neuroimmunol 1998; 83:19-28.

46. House RV, Thomas PT, Bhargava HM. A comparative study of immuno-modulation produced by in vitro exposure to delta opioid receptor agonist peptides. Peptides 1996; 17:75-81.
47. Kay N, Allen J, Morley JE. Endorphins stimulate normal human peripheral blood lymphocyte natural killer activity. Life Sci 1984; 35:53-59.
48. Puente J, Maturana P, Miranda D et al. Enhancement of human natural killer cell activity by opioid peptides: similar response to methionine-enkephalin and beta-endorphin. Brain Behav Immun 1992; 6:32-9.
49. Oleson DR, Johnson DR. Regulation of human natural cytotoxicity by enkephalins and selective opiate agonists. Brain Behav Immun 1988; 2:171-86.
50. Bajpai K, Singh VK, Agarwal SS et al. Immunomodulatory activity of met-enkephalin and its two potent analogs. Int J Immunopharmacol 1995; 17:207-212.
51. Gilman SC, Schwartz JM, Milner RJ et al. beta-Endorphin enhances lymphocyte proliferative responses. Proc Natl Acad Sci USA 1982; 79:226-30.
52. Gilmore W, Weiner LP. The opioid specificity of beta-endorphin enhancement of murine lymphocyte proliferation. Immunopharmacol 1989; 17:19-30.
53. Hemmick LM, Bidlack JM. Beta-endorphin stimulates rat T lymphocyte proliferation. J Neuroimmunol 1990; 29:239-48.
54. Dubinin KV, Zakharova LA, Khegai LA et al. Immunomodulating effect of met-enkephalin on different stages of lymphocyte proliferation induced with concanavalin-A in vitro. Immunopharmacol Immunotoxicol 1994; 16:463-472.
55. Hucklebridge FH, Hudspith BN, Lydyard PM et al. Stimulation of human peripheral lymphocytes by methionine enkephalin and delta-selective opioid analogues. Immunopharmacol 1990; 19:87-91.
56. Hucklebridge FH, Hudspith BN, Muhamed J et al. Methionine-enkephalin stimulates in vitro proliferation of human peripheral lympho- cytes via delta-opioid receptors. Brain Behav Immun 1989; 3:183-9.
57. Foris G, Medgyesi GA, Hauck M. Bidirectional effects of met-enkephalin on macrophage effector functions. Mol Cell Biochem 1986; 69:127-137.
58. Marotti T, Burek B, Rabatic S et al. Modulation of lipopolysaccharide-induced production of cytokines by methionine-enkephalin. Immunol Lett 1994; 40:43-47.
59. Szabo I, Rojavin M, Bussiere JL et al. Suppression of peritoneal macrophage phagocytosis of Candida albicans by opioids. J Pharmacol Exp Ther 1993; 267:703-6.
60. Foris G, Medgyesi GA, Gyimesi E et al. Met-enkephalin induced alterations of macrophage functions. Mol Immunol 1984; 21:747-50.
61. Tosk JM, Grim JR, Kinback KM et al. Modulation of chemiluminescence in a murine macrophage cell line by neuroendocrine hormones. Int J Immunopharmacol 1993; 15:615-20.
62. Sharp BM, Keane WF, Suh HJ et al. Opioid peptides rapidly stimulate superoxide production by human polymoprhonuclear leukocytes and macrophages. Endocrinology 1985; 117:793-5.
63. Radulovic J, Dimitrijevic M, Laban O et al. Effect of met-enkephalin and opioid antagonists on rat macrophages. Peptides 1995; 16:1209-13.
64. Efanov AM, Koshkin AA, Sazanov LA et al. Inhibition of the respiratory burst in mouse macrophages by ultra-low doses of an opioid peptide is consistent with a possible adaptation mechanism. FEBS Lett 1994; 355:114-8.
65. Heijnen CJ, Kavelaars A, Ballieux RE. Beta-endorphin: cytokine and neuropeptide. Immunol Rev 1991; 119:41-63.
66. Panerai AE, Manfredi B, Granucci F et al. The β-endorphin inhibition of mitogen-induced splenocytes proliferation is mediated by central and peripheral paracrine/autocrine effects of the opioid. J Neuroimmunol 1995: 58:71-76.
67. Schneider GM, Lysle DT. Effects of centrally administered opioid agonists on macrophage nitric oxide production and splenic lymphocyte proliferation. In: Friedman H, ed. AIDS, Drugs of Abuse and the Neuroimmune Axis. New York: Plenum Press, 1996.
68. Manfredi B, Sacerdote P, Bianchi M et al. Evidence for an opioid inhibitory effect on T cell proliferation. J Neuroimmunol 1993; 44:43-48.
69. Lysle DT, How T. Endogenous opioids regulate the expression of inducible nitric oxide synthase by splenocytes. J Pharmacol Exp Ther 1999; 288:502-508.
70. Tian M, Broxmeyer HE, Fan Y et al. Altered hematopoiesis, behavior and sexual function in mu opioid receptor deficient mice. J Exp Med 1997; 185:1517-1522.
71. Kieffer BL. Opioids: first lesson from knock-out mice. TiPS 1999; 20:19-25.

72. Mosman TR, Sad S. The expanding universe of T-cell subsets; Th1,Th2 and more. Immunol Today 1996; 17:138-146.
73. Romagnani S. Lymphokine production by human T cell in disease states. Annu Rev Immunol 1994; 12:227-230.
74. Sacerdote P, Manfredi B, Gaspani L et al. The opioid antagonist naloxone induces a shift from type 2 cytokine pattern to type 1 cytokine pattern in balb/cJ mice. Blood 2000; 95:2031-2036.
75. Panerai AE, Sacerdote P. Beta-endorphin in the immune system: a role at last? Immunol Today 1997; 18:317-319.
76. Strom T, Roy-Chaundhury P, Manfro R et al. The Th1/Th2 paradigm and the allograft response. Curr Op Immunol 1996; 8:688-93.
77. Sacerdote P, Rosso di San Secondo VEM, Sirchia G et al. Endogenous opioids modulate allograft rejection time in mice: possible relation with Th1/Th2 cytokines. Clin Exp Immunol 1998; 113:465-469.
78. Panerai AE, Radulovic J, Monastra G et al. Beta-endorphin concentrations in brain areas and peritoneal macrophages in rats susceptible and resistant to experimental allergic encephalomyelitis: a possible relationship between tumor necrosis factor-α and opioids in the disease. J Neuroimmunol 1994; 51:169-76.
79. Sacerdote P, Bianchi M, Panerai AE. Involvement of Beta-endorphin in the modulation of paw inflammatory edema in the rat. Reg Peptides 1996; 63:79-83.
80. Khansari DN, Murgo AJ, Faith RE. Effects of stress on the immune system. Immunol Today 1990; 11:170-5.
81. Shavit Y, Lewis JW, Terman GW et al. Opioid peptides mediate the suppressive effects of stress on natural killer cell cytotoxicity. Science 1984; 223:188-190.
82. Sacerdote P, Manfredi B, Bianchi M et al. Intermittent but not continous inescapable footshock stress affects immune responses and and immunocyte beta-endorphin concentrations in the rat. Brain Behav Immunity 1994; 8:251-60.
83. Sacerdote P. Interleukins and immunocyte beta-endorphin. In: Plotnikoff, Faith, Murgo, Good, eds. Cytokines, Stress and Immunity. CRC Press, 1999:271-280.

CHAPTER 10

The Immune-Suppressive Effects of Pain

Gayle G. Page

Introduction

The immune-suppressive effects of painful experiences have been studied in both humans and animals for many years. Experimental pain has been induced by such means as electric shock and surgery in animals, and humans undergoing surgery have been studied extensively. In general, results have shown such perturbations to suppress the immune functions that are assessed.[1,2] The exclusive contribution of the pain per se to these findings has only recently become a focus of study. If pain mechanisms were shown to mediate the observed immunosuppressive effects of experiences such as recovery from injury or undergoing surgery, then adequate pain management would become a vital adjunct to the overall care of such individuals.

The importance of such an avenue of study relates to the crucial role played by the immune system in maintaining health and resisting infection and disease.[3] In the latter case, animal studies provide direct evidence that natural killer (NK) cells play a key role in controlling metastatic processes[4,5] as well as support for the hypothesis that the suppressed NK cell activity following surgery underlies surgery-induced metastatic promotion.[1,6] Findings of human studies are corroborative; low NK activity during the perioperative period is associated with higher rates of cancer recurrence and mortality in patients with breast[7], head and neck[8], lung[9], and colorectal cancers.[10]

Animal Studies of Pain and Immune Suppression

Aversive stressors such as footshock and tail shock have been used as means by which to provoke pain without tissue damage in rats. Intermittent inescapable footshock delivered over some minutes has been shown to suppress several immune functions, including NK cell activity,[2,11] peripheral blood mononuclear or spleen cell proliferative responses to mitogens including phytohemagglutinin and Concavalin A.[2,12] Similarly, intermittent tail shock has been shown to suppress mixed lymphocyte reactions in lymph node cells[13] and reduce in vivo responses to a novel antigen assessed as immunoglobulin (Ig) G antibody levels.[14] An important consideration in these studies is the psychological stress that likely occurs in animals subjected to some period of repeated uncontrollable aversive stimuli;[15,16] thus, the observed immune suppression may be attributed to mechanisms other than pain per se.

Experimental surgery has been shown to suppress immune functions and to promote tumor development in rats and mice. Rats recovering from laparotomy exhibit decreases in both lymphocyte and splenocyte proliferative responses to mitogens and in NK cell

Immune Mechanisms of Pain and Analgesia, edited by Halina Machelska and Christoph Stein.

activity;[17-21] more invasive surgery is associated with greater decrements in immune function.[18,19] Similarly, undergoing surgery promotes tumor development and the magnitude of tumor promotion is associated with the invasiveness of surgery.[6,22,23]

Pain and Immune Function in Humans

In humans, surgery is well known to result in immune suppression, including lymphocyte proliferative responses to mitogens and NK cell activity,[24-27] decreased cell mediated immunity,[28,29] and alterations in the balance of Th1 versus Th2 cells, compromising cellular immunity.[30,31] The invasiveness of the surgery has been associated with the magnitude of immune suppression,[31] and immune function is comparatively more suppressed by surgery in individuals with cancer,[29] although this finding is not a consistent one.[32] If pain is a mediator of such surgery-induced immune suppression, then it would seem that anesthesia and analgesia techniques that reduce perioperative pain would also impact the observed postoperative immune suppression. Indeed, several studies comparing general inhalational anesthesia versus epidural or spinal anesthesia for surgical procedures that are below the waist have shown that immune outcomes were significantly better preserved in groups treated with epidural anesthesia, including NK cell activity following laparotomy[33] and hysterectomy;[34] however, this benefit was not observed in patients undergoing total hip replacement.[25] Given the importance of immune function in resisting infection, that epidural anesthesia was also shown to result in fewer infectious complications compared to standard inhalational anesthesia[35] supports the above-mentioned suggestion.

Interleukin (IL)-6, a cytokine produced by a host of immune cells including monocytes, macrophages and lymphocytes, is involved in a broad array of actions including inflammation and the regulation of endocrine and metabolic functions, notably the hypothalamic-pituitary-adrenal axis and catecholamines.[36] Undergoing surgery results in increased IL-6 levels in plasma[37] as well as cerebrospinal fluid.[38] Several direct comparisons have shown that more invasive surgeries are associated with greater levels of plasma IL-6, including laparoscope assisted vaginal versus abdominal hysterectomy[39], and cholecystectomy using a laparoscope versus an open abdominal approach.[40,41] Kristiansson et al[41] also showed the laparoscopic approach resulted in reduced pain scores. Given that prostaglandins promote inflammatory changes and hyperalgesia,[42] and that prostaglandins have been associated with NK suppression,[43,44] interventions that reduce inflammation would be expected to both reduce pain and its immunosuppressive effects. In animals, prostaglandin E_2 antagonism via cyclooxygenase inhibition or a monoclonal antibody reduced paw swelling, and both paw and serum IL-6 levels resulting from adjuvant arthritis or carrageenan injection,[42,45] and significantly reduced thermal hyperalgesia.[42] In humans undergoing cholecystectomy, perioperative ibuprofen administration resulted in significantly lower levels of plasma IL-6 levels compared to placebo controls.[37]

Animal Studies of Pain, Metastasis and Immune Suppression

Fig. 1 offers selected possible interactions among nociceptive processes, local metabolic processes, neuroendocrine activation and immunity that are altered as a result of the cutting, tearing, and manipulation of tissues in the conduct of surgery. With particular relevance to the studies to be discussed are hypothalamic activation by ascending nociceptive (painful) impulses, and the resulting increases in sympathetic nervous system activity with epinephrine release and corticosteroid release from the adrenals, both of which have been shown to suppress NK cytotoxic activity.[46-48] The release of local factors from the

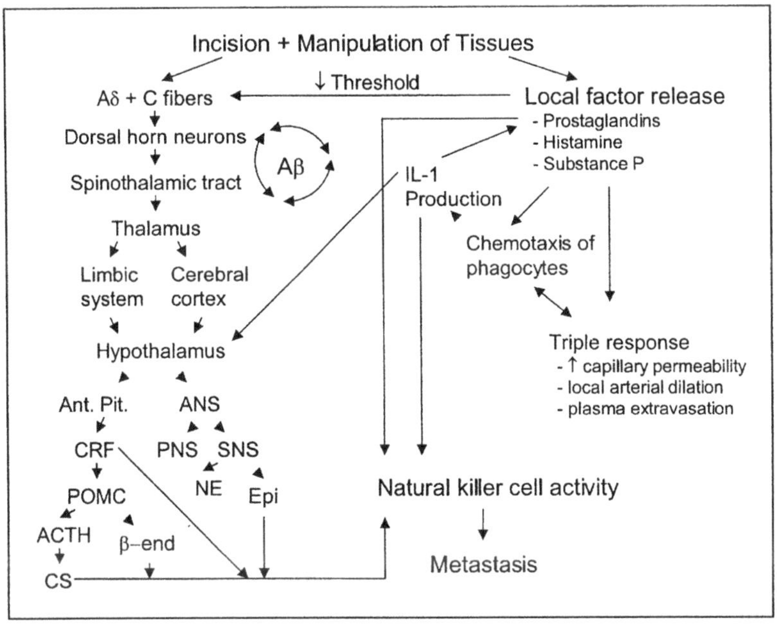

Fig. 1. Interactions among nociceptive and local metabolic processes, neuroendocrine activation and immunity, ultimately impacting resistance against metastasis. Aδ, Aβ and C refer to peripheral nerve fiber types. Abbreviations: IL-1,interleukin-1; Ant. Pit., anterior pituitary; ANS, autonomic nervous system; CRF, corticotropin releasing factor; PNS, parasympathetic nervous system; SNS, sympathetic nervous system; POMC, pro-opiomelanocortin; NE, norepinephrine; Epi, epinephrine; ACTH, adrenocorticotropic hormone; β-end, β-endorphin; CS, corticosterone.

injury site further facilitate central pain processing by sensitizing peripheral fibers, thereby reducing the threshold for nociceptive impulse transmission and the self-perpetuating hyper-responsiveness to mechanical and thermal stimuli at the surgical site.[49] Taken together, both central and peripheral changes contribute to the suppression of NK cytotoxicity that is observed postoperatively.

To attribute some measure of biologic significance to NK suppression, we have used an NK-sensitive mammary adenocarcinoma cell line, MADB106, syngeneic to the inbred Fischer 344 rats. MADB106 tumor cells seed and colonize only in the lungs following intravenous injection, and the NK sensitivity of these processes has been shown to be limited to the first 24 hours after injection in a time dependent and decremental manner.[5,50] Thus, both the colonization of MADB106 cells in the lungs assessed at 3 weeks after tumor injection and the lung retention of radiolabeled MADB106 cells assessed at 18-24 hours after tumor injection provides an indication of host susceptibility to metastasis as well as in vivo levels of NK activity throughout the early hours following tumor injection.

The overall goal of our work is to investigate possible mediation of specific pain mechanisms in surgery-induced decreases in NK activity and host resistance against metastasis. To explore the possibility that pain per se mediates these negative consequences of surgery, pharmacologic interventions have been employed with the rationale that if the drug significantly attenuates surgery-induced reductions in both host resistance against metastasis and exploratory behavior, then pain relief might be suggested as a biologically significant and beneficial treatment.

Animals

Mature Fischer 344 rats were used in all studies. Animals were maintained for 4 weeks following shipment to allow for acclimatization to the vivarium and the 12-hour dark/light schedule. Animals had unlimited access to food and water until 8 hours prior to surgery, when only water was available. All surgeries were performed within the first 5 hours of the dark phase. All protocols were approved by the institutional committee for the care and use of animal subjects.

Surgery

Animals were anesthetized with halothane throughout the preparation, including skin preparation and antibiotics injection, and conduct of the surgery. The experimental laparotomy consisted of a 4 cm midline incision with externalization of 10 cm of the small intestine, followed by gentle rubbing of the intestine between two pieces of gauze at 4 different locations. The intestine remained exteriorized for a total of 4 minutes under saline-soaked gauze, and was then returned to the abdominal cavity, irrigated, and the muscle and skin layers sutured with monofilament wire. "Anesthesia only animals" were anesthetized at the same time and in the same dose as the surgery animals.

The Beneficial Effects of Morphine on Surgery-Induced Decreases in Host Resistance Against Metastasis

The first studies investigating the possibility that providing an analgesic dose of morphine would reduce the metastatic-enhancing effects of surgery used a simple 2×2 experimental design: Surgery versus Anesthesia by Morphine versus Vehicle. Pre- and postoperative morphine administration was accomplished in two different preparations: in saline intraperitoneally 30 minutes before surgery to insure that surgery was conducted at the time of peak analgesic effect, and in a slow-release suspension (SRS) subcutaneously, an oil emulsion that releases the drug over some hours, for drug administrations immediately following and at 5 hours postoperative. As for all studies discussed herein, syngeneic NK-sensitive MADB106 tumor cells were injected at 5 hours after surgery in an effort to capture the stressful nature of the postoperative experience rather than responses to intraoperative events.

We found a significant interaction between the effects of surgery and morphine such that morphine attenuated the observed surgery-induced increase in both the colonization of MADB106 cells in the lungs assessed at 3 weeks after surgery[51] and the retention of radiolabeled MADB106 cells in the lungs at 18 hours postoperative.[52] Surgery resulted in a more than 5-fold increase in lung tumor cell retention and morphine treatment reduced this effect by more than 50%.[52] Morphine administration exerted no significant effects in the "anesthesia only animals" for both tumor outcomes.[51,52]

Exploratory behavior studies have been an important aspect of our work. The rationale for such observations relates to the clinical behavior of humans following major abdominal surgery, the characteristic splinting of the abdomen and hunching posture. It was noted that following abdominal surgery, rats also exhibit such a posture and are hesitant to rear (lift both forepaws off of the cage floor), thus stretching the incision in a way not unlike the human assuming an erect posture. Therefore, for exploratory behavior studies, the number of rears exhibited by each animal during the latter 30 minutes of each of the first 4 postoperative hours was enumerated to serve as an indicator of abdominal discomfort. Surgery resulted in a profound suppression of rearing behavior and morphine administration restored rearing incidences to the levels observed in the unoperated rats.[51]

To further investigate the possibility that it was the pain-relieving effects of morphine that were responsible for its beneficial effect on the NK-suppressive and metastatic-enhancing consequences of surgery, we designed a subsequent study to assess the relative importance of the pre- versus postoperative administration of morphine in this paradigm. Either morphine or its vehicle was administered to all animals at the same three above-mentioned time points: (1) 30 minutes before surgery, (2) at the completion of surgery in SRS, and (3) at the time of radiolabeled MADB106 tumor cell injection at 5 hours after surgery in SRS. Surgery groups received either the vehicle or morphine in one of four different regimens: before surgery only, at all three times, after surgery only at times 2 and 3, and after surgery total at times 2 and 3 with the preoperative dose added at time 2. Two anesthesia control groups received either morphine or vehicle at all three times. Surgery resulted in a two-fold increase in lung tumor retention, which was significantly attenuated by all four morphine treatment regimens. Furthermore, the two surgery groups that were treated with morphine preoperatively appeared to derive greater benefit, suggesting that the preoperative administration of an opioid is key in optimizing its beneficial effects on surgery-induced increases in metastatic susceptibility. With regard to exploratory behavior, morphine treatment similarly restored surgery-induced activity suppression in all animals except those that received both the pre- and postoperative morphine doses at the completion of surgery. These animals were laying on their bellies, virtually unmoving until the third postoperative hour, likely indicating over-medication.[53]

Pre- and Postoperative Fentanyl Attenuates Surgery-Induced Increases in Metastatic Susceptibility

The more selective μ-agonist, fentanyl, was used to further explore the possibility that the pain per se is a mediator of surgery-induced increases in metastatic susceptibility. Additionally, female rats were used in this and the subsequent studies described herein. Using the same 2×2 design of surgery versus anesthesia by drug versus vehicle, fentanyl citrate was administered subcutaneously 30 minutes preoperatively, and in the SRS immediately after the completion of surgery. Surgery resulted in a more than four-fold increase in the lung tumor retention of radiolabeled MADB106 cells, which was significantly ameliorated by fentanyl administration; fentanyl exerted no effects in the "anesthesia only animals". Importantly, females derived similar benefits from drug treatment.

With regard to behavioral activity, operated animals exhibited a significant reduction in rearing behavior compared to the nonoperated animals. Fentanyl treatment did not restore activity levels to those of the unoperated animals; however, among the surgery animals, fentanyl treatment resulted in a significant increase in activity levels compared to the vehicle-injected animals.[54]

The Beneficial Effects of Preoperative Intrathecal Bupivacaine Plus Morphine on Surgery-Induced Decreases in Host Resistance Against Metastasis

This study explored the possibility that blocking ascending impulses from the abdominal laparotomy would affect the tumor-enhancing effects of surgery. Thus, the same above-mentioned 2×2 design was employed and drug treated animals received an intrathecal (between L4 and L5 lumbar vertebrae) injection of bupivacaine combined with morphine immediately before the conduct of surgery. Control animals received intrathecal saline injections. Surgery resulted in a more than three-fold increase in the lung tumor retention of radiolabeled MADB106 cells, which was significantly attenuated by the intrathecal injection of bupivacaine plus morphine; intrathecal drug injections exerted no effects in the "anesthesia only animals". Females and males derived equivalent benefits from the drug intervention.[54] In a separate set of experiments, the preoperative intrathecal administration of similar doses of bupivacaine plus morphine significantly reduced surgery-induced increases in both the lung retention of radiolabeled MADB106 cells and the number of tumor colonies evident three weeks after MADB106 tumor injection.[55]

Postoperative Indomethacin Administration Attenuates Surgery-Induced Increases in Host Susceptibility to Metastasis

Given the contribution of prostaglandins to wound inflammatory changes and hyperalgesia, the cyclooxygenase inhibitor indomethacin was employed as the drug treatment for this study. Using the same 2×2 design, drug treated animals received a subcutaneous injection of indomethacin at the completion of surgery. Surgery resulted in a more than two-fold increase in the lung tumor retention of radiolabeled MADB106 cells, which was significantly attenuated by indomethacin treatment in both males and females. Indomethacin administration exerted no effects in the "anesthesia only animals". This significant interaction between the effects of surgery and drug treatment were also observed in the animals' activity levels. Specifically, indomethacin administration restored rearing incidence to normal levels in both the males and females.[56]

In a subsequent study, at the time of tumor injection, 5 hours postoperative, all animals underwent cardiac blood withdrawal for the assessment of plasma levels of IL-6 by quantitative sandwich enzyme immunoassay (ELISA). We found that no "anesthesia only animal" exhibited detectable plasma levels of IL-6. Among the surgery animals, indomethacin administration resulted in a significant reduction of surgery-induced increases in plasma IL-6 levels in the males, no such benefits were observed in the females (unpublished data).

Discussion and Conclusions

Taken together the findings of these studies support the suggestion that the provision of pain relief ameliorates surgery-induced decreases in host resistance against metastasis. In particular, these studies show that a variety of morphine regimens provide protection against the tumor-enhancing effects of surgery including both pre- and postoperative administration[51,52] as well as only preoperative or only postoperative treatment.[53] The pre- and postoperative systemic administration of fentanyl confers similar benefits as morphine in this paradigm and is also beneficial in the females.[54] The benefits conferred by the preoperative intrathecal injection of bupivacaine plus morphine[54] as well as the postoperative systemic administration of indomethacin[56] on the metastatic-enhancing effects of surgery further support the suggestion that pain is a mediator of this negative consequence of undergoing and recovering from surgery. That either morphine or indomethacin administration completely restored surgery-induced activity suppression to normal levels, and that fentanyl significantly improved surgery-induced activity suppression suggests that these regimens provided pain relief.

Two points support the hypothesis that it is the pain relieving effects of these drugs that attenuated surgery-induced tumor promotion rather than direct effects on immunity, the tumor cells themselves, or other mechanisms that affect metastasis. First, the very low doses and intrathecal administration of bupivacaine plus morphine argues against possible direct peripheral effects of these drugs. Second, all of the analgesia regimens provided benefit only to the animals undergoing surgery; there were no significant drug-induced changes in metastatic susceptibility in animals not experiencing pain. Such an interaction between the effects of the analgesic agents and exposure to surgery implicates some blockade of the impact of surgery rather than an independent drug effect on tumor development.

In conclusion, these findings show that some analgesic techniques provide significant protection against the tumor-enhancing effects of undergoing and recovering from surgery. Although it remains unknown whether the alleviation of perioperative pain will provide similar benefits in humans, the current studies suggest that it will, specifically in patients with potentially metastasizing cancer. Therefore, it is suggested that the relief of perioperative pain becomes a high priority in the care of individuals with cancer.

References

1. Ben-Eliyahu S, Page GG, Yirmiya R et al. Evidence that stress and surgical interventions promote tumor development by suppressing natural killer cell activity. Int J Cancer 1999; 80:880-888.
2. Sacerdote P, Manfredi B, Bianchi M et al. Intermittent but not continuous inescapable footshock stress affects immune responses and immunocyte beta-endorphin concentrations in the rat. Brain Behav Immun 1994; 8:251-260.
3. Whiteside TL, Herberman RB. Role of human natural killer cells in health and disease. Clin Diagn Lab Immunol 1994; 1:125-133.
4. Wiltrout RH, Herberman RB, Zhang S et al. Role of organ-associated NK cells in decreased formation of experimental metastases in lung and liver. J Immunol 1985; 134:4267-4275.
5. Ben-Eliyahu S, Page GG. The in vivo assessment of natural killer cell activity in rats. Prog Neuro Endocrin Immunol 1992; 5:199-214.
6. DaCosta ML, Redmond HR, Bouchier-Hayes DJ. The effect of laparotomy and laparoscopy on the establishment of spontaneous tumor metastasis. Surgery 1998; 124(3):516-525.
7. Levy SM, Herberman RB, Maluish AM et al. Prognostic risk assessment in primary breast cancer by behavioral and immunological parameters. Health Psychol 1985; 4:99-113.
8. Schantz SP, Taylor DL, Racz T et al. Natural killer cells and metastases from pharyngeal carcinoma. Am J Surg 1989; 158:361-366.
9. Fujisawa T, Yamaguchi Y. Autologous tumor killing activity as a prognostic factor in primary resected nonsmall cell carcinoma of the lung. Cytopathology 1997; 79:474-481.
10. Koda K, Saito N, Takiguchi N et al. Preoperative natural killer cell activity: Correlation with distant metastases in curatively research colorectal carcinomas. Int Surg 1997; 82:190-193.
11. Shavit Y, Martin FC, Yirmiya R et al. Effects of single administration of morphine or footshock stress on natural killer cell cytotoxicity. Brain Behav Immun 1987; 1:318-328.
12. Pezzone MA, Dohanics J, Rabin BS. Effects of footshock stress upon spleen and peripheral blood lymphocyte mitogenic responses in rats with lesions of the paraventricular nuclei. J Neuroimmunol 1994; 53:39-46.
13. Fleshner M, Bellgrau D, Watkins LR et al. Stress-induced reduction in the rat mixed lymphocyte reaction is due to macrophages and not to changes in T cell phenotypes. J Neuroimmunol 1995; 56:45-52.
14. Laudenslager ML, Fleshner M, Hofstadter P et al. Suppression of specific antibody production by inescapable shock: Stability under varying conditions. Brain Behav Immun 1988; 2:92-101.
15. Kant GJ, Bauman RA, Anderson SM et al. Effects of controllable vs. uncontrollable chronic stress on stress-responsive plasma hormones. Physiol Behav 1992; 51:1285-1288.
16. Drugan RC, Basile AS, Ha JH et al. Analysis of the importance of controllable versus uncontrollable stress on subsequent behavioral and physiological functioning. Brain Res Brain Res Protoc 1997; 2:69-74.
17. Nelson CJ, Lysle DT. Severity, time, and beta-adrenergic receptor involvement in surgery-induced immune alterations. J Surg Res 1998; 80(2):115-122.
18. Toge T, Hirai T, Takiyama W et al. Effects of surgical stress on natural killer activity, proliferative response of spleen cells and cytostatic activity of lung macrophages in rats. Gann 1981; 72:790-794.
19. Sandoval BA, Robinson AV, Sulaiman TT et al. Open versus laparoscopic surgery: A comparison of natural antitumoral cellular immunity in a small animal model. Am Surg 1996; 62(8):625-631.

20. Colacchio TA, Yeager MP, Hildebrandt L. Perioperative immunomodulation in cancer surgery. Am J Surg 1994; 167:174-179.
21. Pollock RE, Lotzová E, Stanford SD et al. Effect of surgical stress on murine natural killer cell cytotoxicity. J Immunol 1987; 138:171-178.
22. Allendorf JD, Bessler H, Horvath KD et al. Increased tumor establishment and growth after open vs laparoscopic bowel resection in mice. Surg Endosc 1998; 12:1035-1038.
23. Lee SW, Southall JC, Allendorf JD et al. Tumor proliferative index is higher in mice undergoing laparotomy vs. CO2 pneumoperitoneum. Dis Colon Rectum 1999; 42:477-481.
24. Sacerdote P, Bianchi M, Gaspani L et al. The effects of tramadol and morphine on immune responses after surgery in cancer patients. Anesth Analg 2000; 90(6):1411-1414.
25. Salo M, Nissilä M. Cell-mediated and humoral immune responses to total hip replacement under spinal or general anaesthesia. Acta Anaesthesiol Scand 1990; 34:241-248.
26. Kutza J, Gratz I, Afshar M et al. The effects of general anesthesia and surgery on basal and interferon stimulated natural killer cell activity of humans. Anesth Analg 1997; 85(4):918-923.
27. Pollock RE, Lotzová E, Stanford SD. Mechanism of surgical stress impairment of human perioperative natural killer cell cytotoxicity. Arch Surg 1991; 126:338-342.
28. Faist E, Ertel W, Cohnert T et al. Immunoprotective effects of cyclooxygenase inhibition in patients with major surgical trauma. J Trauma 1990; 30:8-18.
29. Riboli EB, Terrizzi A, Arnulfo G et al. Immunosuppressive effect of surgery evaluated by the multitest cell-mediated immunity system. Can J Surg 1984; 27(1):60-63.
30. Le Cras AE, Galley HF, Webster NR. Spinal but not general anesthesia increases the ratio of T helper 1 to T helper 2 cell subsets in patients undergoing transurethral resection of the prostate. Anesth Analg 1998; 87:1421-1425.
31. Decker D, Schöndorf M, Bidlingmaier F et al. Surgical stress induces a shift in the type-1/type-2 T-helper cell balance, suggesting down-regulation of cell-mediated and up-regulation of antibody-mediated immunity commensurate to trauma. Surgery 1996; 119:316-325.
32. Beilin B, Shavit Y, Hart J et al. Effects of anesthesia based on large versus small doses of fentanyl on natural killer cell cytotoxicity in the perioperative period. Anesth Analg 1996; 82:492-497.
33. Koltun WA, Bloomer MM, Tilberg AF et al. Awake epidural anesthesia is associated with improved natural killer cell cytotoxicity and reduced stress response. Am J Surg 1996; 171:68-73.
34. Tønnesen E, Wahlgreen C. Influence of extradural and general anaesthesia on natural killer cell activity and lymphocyte subpopulations in patients undergoing hysterectomy. Br J Anaesth 1988; 60:500-507.
35. Yeager MP, Glass DD, Neff RK et al. Epidural anesthesia and analgesia in high-risk surgical patients. Anesthesiology 1987; 66:729-736.
36. Papanicolaou DA, Wilder RL, Manolagas SC et al. The pathophysiologic roles of interleukin-6 in human disease. Ann Int Med 1998; 128:127-137.
37. Chambrier C, Chassard D, Bienvenu J et al. Cytokine and hormonal changes after cholecystectomy. Effect of ibuprofen pretreatment. Ann Surg 1996; 224(2):178-182.
38. Yeager MP, Lunt P, Arruda J et al. Cerebrospinal fluid cytokine levels after surgery with spinal or general anesthesia. Reg Anesth Pain Med 1999; 25(6):557-562.
39. Labib M, Palfrey S, Paniagua E et al. The postoperative inflammatory response to injury following laparoscopic assisted vaginal hysterectomy versus abdominal hysterectomy. Ann Clin Biochem 1997; 34:543-545.
40. Bellón JM, Manzano L, Larrad A et al. Endocrine and immune response to injury after open and laparoscopic cholecystectomy. Int Surg 1998; 83:24-27.
41. Kristiansson M, Saraste L, Soop M et al. Diminished interleukin-6 and C-reactive protein responses to laparoscopic versus open cholecystectomy. Acta Anaesthesiol Scand 1999; 43:146-152.
42. Portanova JP, Zhang Y, Anderson GD et al. Selective neutralization of prostaglandin E_2 blocks inflammation, hyperalgesia, and interleukin–6 production in vivo. J Exp Med 1996; 184:883-891.
43. Ellis NK, Duffie GP, Young MR et al. The effects of 16,16-dimethyl PGE2 and phosphodiesterase inhibitors on con A blastogenic responses and NK cytotoxic activity of mouse spleen cells. J Leukoc Biol 1990; 47:371-377.
44. Lala PK, Parhar RS, Singh P. Indomethacin therapy abrogates the prostaglandin-mediated suppression of natural killer activity in tumor-bearing mice and prevents tumor metastasis. Cell Immunol 1986; 99:108-118.

45. Anderson GD, Hauser SD, McGarity KL et al. Selective inhibition of cyclooxygenase (COX)-2 reverses inflammation and expression of COX-2 and interleukin-6 in rat adjuvant arthritis. J Clin Invest 1996; 97(11):2672-2679.
46. Ben-Eliyahu S, Shakhar G, Page GG et al. Suppression of NK cell activity and of resistance to metastasis by stress: A role for adrenal catecholamines and β-adrenoceptors. Neuroimmunomodulation, in press.
47. Shakhar G, Ben-Eliyahu S. In vivo beta-adrenergic stimulation suppresses natural killer activity and compromises resistance to tumor metastasis in rats. J Immunol 1998; 160(7):3251-3258.
48. Gatti G, Cavallo R, Sartori ML et al. Cortisol at physiological concentrations and prostaglandin E_2 are additive inhibitors of human natural killer cell activity. Immunopharmacology 1986; 11(2):119-128.
49. Woolf CJ, Salter MW. Neuronal plasticity: Increasing the gain in pain. Science 2000; 288:1765-1768.
50. Barlozzari T, Leonhardt J, Wiltrout RH et al. Direct evidence for the role of LGL in the inhibition of experimental tumor metastases. J Immunol 1985; 134:2783-2789.
51. Page GG, Ben-Eliyahu S, Yirmiya R et al. Morphine attenuates surgery-induced enhancement of metastatic colonization in rats. Pain 1993; 54:21-28.
52. Page GG, Ben-Eliyahu S, Liebeskind JC. The roll of LGL/NK cells in surgery-induced promotion of metastasis and its attenuation by morphine. Brain Behav Immun 1994; 8:241-250.
53. Page GG, McDonald JS, Ben-Eliyahu S. Preoperative versus postoperative administration of morphine: Impact on the neuroendocrine, behavioural, and metastatic-enhancing effects of surgery. Br J Anaesth 1998; 81:216-223.
54. Page GG, Blakely WP, Ben-Eliyahu S. Evidence that postoperative pain is a mediator of the tumor-promoting effects of surgery in rats. Pain 2001; 90:191-199.
55. Bar-Yosef S, Melamed R, Page GG et al. Attenuation of the tumor-promoting effect of surgery by spinal blockade in rats. Anesthesiology 2001; 94:1066-1073.
56. Page GG, Blakely WP, Duong THT. Evidence that prostanoids contribute to surgery-induced NK suppression and tumor promotion. Soc Neurosci Abstr 2000; 26.

Chapter 11

Invertebrate Opiate Immune and Neural Signaling

George B. Stefano, Patrick Cadet,
Christos M. Rialas, Kirk Mantione, Federico Casares,
Yannick Goumon and Wei Zhu

Introduction

It is increasingly becoming evident that specific chemical signal molecules within a cell and between cells can be the same regardless of whether an organism is an invertebrate or vertebrate. This is true for monoaminergic, opioid, opiate and endocannabinoid signal molecules, just to name a few (see refs. 1-4). It would appear that a mechanism exists to maintain these signaling families during evolution. In this regard, we have previously proposed that this signaling is conserved during evolution due to the numerous conformational matching processes found in a signaling system (enzymatic synthesis, degradation and receptor interaction).[1,5,6] Thus, stereospecificity imposes rigidity on the development of new signal molecules since so many conformational matches must occur and manifest themselves simultaneously. It comes as no surprise to find these important communication molecules, i.e., morphine, throughout the animal kingdom, and some in the plant kingdom as well, indicating that they may have their origins even further back in evolution, sharing a common ancestor.

This chapter will focus on opioid processes specifically in the marine bivalve mollusk, *Mytilus edulis,* since it has served as a continuous source of information concerning comparative opioid processes. Initially, this animal was chosen as this type of model because it contained monoaminergic peripheral and central nervous system (CNS) signaling.[1] Additionally, it is a long-lived invertebrate, living for up to 17 years in the Long Island environment.[7] In this regard, it was surmised 30 years ago that this particular organism would have to depend on established and sophisticated neural processes to cope with its environment, as opposed to surviving as a result of a huge reproductive effort. On the other hand, short-lived invertebrates may not require a large and relatively sophisticated nervous or immune system, since they can survive to their reproductive time via producing greater numbers of animals, thus ensuring their species survival. This demonstrates, for example, that neural and immune capacities truly function to provide longevity. Therefore, longevity can be enhanced when these systems communicate.

Immune Mechanisms of Pain and Analgesia, edited by Halina Machelska and Christoph Stein.

The Presence of Opioids and Their Binding Sites

The presence of biologically active neuropeptides in invertebrates, which are comparable to those of vertebrates, has been known for a considerable period of time.[8-11] However, detailed information on a specific class of these peptides, the endogenous opioids, was almost exclusively confined to the mammalian nervous system. The recent upsurge of interest in the diverse roles and modes of operation of these molecules, including immune actions, has sparked a search for their evolutionary history. There have been several reports on the occurrence of endogenous opioids in sub-mammalian vertebrates[12] (for review see ref. 13). Additionally, adrenocorticotropin (ACTH) and β-endorphin-like amino acid sequences were detected immunologically in protozoa as part of a high-molecular-weight macromolecule.[14]

In regard to invertebrates, the presence of an opiate receptor mechanism in the CNS of the marine mollusk, *M. edulis*, was first suggested by a rise in ganglionic dopamine levels following cardiac administration of exogenous met- and leu-enkephalin, an effect reversible by naloxone.[1] The first actual demonstration of high affinity opiate binding sites in an invertebrate ganglion was accomplished by Stefano and colleagues[5] in *M. edulis*. The biochemical characteristics of this system, analyzed in detail by Kream and colleagues,[15] have been found to parallel those of mammalian systems. With respect to specific binding sites in insects, early indications existed.[16,17] It was shown that suspensions prepared from *Drosophila sp* heads avidly bind (^{3}H) leu-enkephalin and the opioid ligand (^{3}H)-diprenorphine. Specific high affinity binding sites for a synthetic enkephalin analog, D-Ala^2-Met^5-enkephalinamide (DAMA), were demonstrated in the cerebral ganglia and mid-gut of the insect *Leucophaea sp*.[18] The results strongly suggest the presence of opiate receptors that are confined to certain areas of nervous tissue in insects. Again these opioid receptors were found to resemble those described in mammalian systems.

Opioid Peptide Precursors

We have found that the difficulty in obtaining these peptides in invertebrate tissues is due to the presence of proteolytic enzymes, i.e., neutral endopeptidase, giving these peptides an extremely short half-life. In *M. edulis* neural tissues, met-enkephalin's half-life is 90 seconds in dissected pedal ganglia unless a "soup" of enzyme inhibitors are present.[19-21] Furthermore, probably due to these peptidases, basal opioid peptide levels in these tissues are always low, complicating detection even more. In retrospect, we now surmise that increased levels of opioid peptides probably arise from their precursors as a result of rapid processing, since they are present (see below). Since this appears to be the case, we can further speculate that the opioid peptides are maintained at these low levels under "basal" conditions to avoid inappropriate signaling. In part, this may be an accommodation to having an "open" circulatory system, which allows for greater cellular interaction without compartments. In this environment it would be especially important to limit and diminish inappropriate signaling. An important strategy, therefore, would be a reliance on enzymatic signal degradation.

Our hypothesis is further supported by the study of signaling peptides in insects. More than 200 peptides were purified from these invertebrates. Most of these signaling molecules were found to be released from brain-copora cadiaca and copora allata complexes (counterpart of human hypothalamus and pituitary) into hemolymph to regulate diverse physiological processes.[22,23] Similar to the opioid peptides in *M. edulis*, the half-life of many insect peptides is very short. The half-life of Neb-Trypsin Modulating Oostatic

Factor (Neb-TMOF) in fly hemolymph is 30 seconds.[24] Other peptides such as Locusta Kinin and Locusta Tachykinin are digested rapidly by peptidase in the hemolymph as well; their half-life is less than 5 min (Wei Zhu, unpublished results). We further found enzymatic degradation of these peptides could be potently inhibited by specific mammalian angiotensin converting enzyme and neutral endopeptidase inhibitors,[24] supporting the opioid peptide processing role as well.[2]

Proenkephalin

A mammalian-like proenkephalin peptide in invertebrates was surmised from studies demonstrating the presence of smaller peptides that are found within this precursor. The opioid peptides met- and leu-enkephalin as well as met-enkephalin-Arg-Phe were isolated and sequenced from *M. edulis* neural tissues,[25,26] arthropods,[27,28] annelids[29] and the mollusk *Lymnaea stagnalis*.[30] Recently, a proenkephalin-like peptide was identified in the immunocytes of two representative invertebrates, namely in the leech *Theromyzon tessulatum*, and in the marine mussel *M. edulis* (Fig. 1).[31] The structure of the leech proenkephalin material demonstrates considerable amino acid sequence similarity with amphibian proenkephalin (26.2%). *M. edulis* proenkephalin exhibits a higher sequence identity with human and guinea pig proenkephalin (39% and 50%, respectively). This proenkephalin contains met- and leu-enkephalin in a ratio of 3:1 for *M. edulis* and 1:2 in the leech. They also possess met-enkephalin-Arg-Gly-Leu and Met-enkephalin-Arg-Phe that are flanked by dibasic amino acid residues, demonstrating cleavage sites. Furthermore, using both sequence comparison and a specific antiserum raised against bovine proenkephalin A (209-237), the enkelytin peptide, FAEPLPSEEEGESYSKEVPEMEKRYGGFM, was identified in invertebrate proenkephalin (Fig. 1) and it exhibited a sequence identity of 98% with mammalian enkelytin.[32]

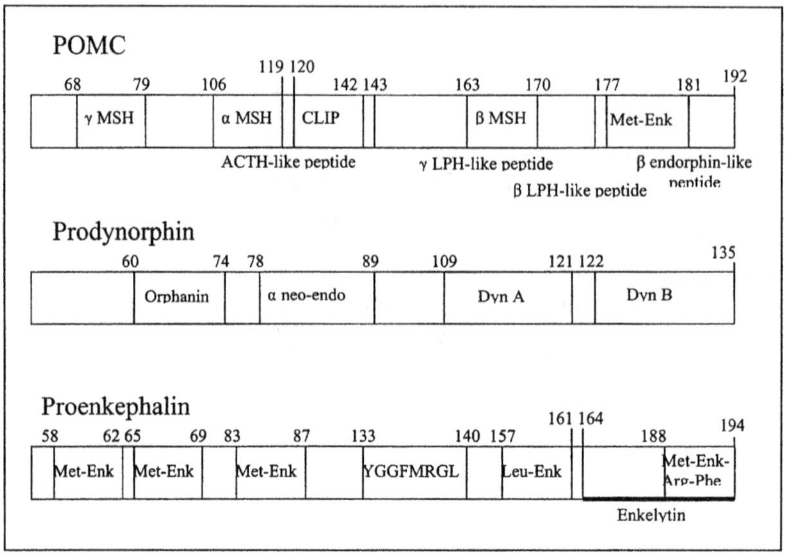

Fig. 1. Peptide gene product illustration is prepared from data published in the following refs. 31,45,52.

The presence of invertebrate (*M. edulis*) enkelytin[32] with a nearly perfect sequence match to that found in bovine chromaffin cells (98%)[32] further supports the hypothesis that these molecules first evolved in simpler animals. Indeed, enkelytin, with its high antibacterial activity[32] further associates opioid peptides with immune related activities.[33] We have recently demonstrated in man and invertebrates that immune signaling or alerting may lead to enhanced proenkephalin proteolytic processing, releasing opioid peptides and enkelytin (Fig. 2).[33-35] In this scenario, the opioid peptides would stimulate immunocyte chemotaxis and phagocytosis as well as the secretion of classical cytokines.[36,37] During this process the liberated enkelytin would simultaneously attack bacteria immediately, thereby allowing time for the immune stimulating capabilities of opioid peptides, i.e., cytokine induction, to manifest itself. This hypothesis is further supported by the presence of specific met-enkephalin receptors on both vertebrate and invertebrate immunocytes.[36,38-40] Interestingly, this same scenario may occur in neural tissues[38,41] given the presence of glial cell types, i.e., microglia. Thus, it appears that many of the mammalian molecular and cellular survival strategies first appeared in organisms that evolved at least 500 million years ago.

Prodynorphin

Alpha-neo-endorphin was purified and sequenced in the leech *T. tessulatum*,[42] suggesting the presence of prodynorphin in invertebrates. Sequence alignment of the entire invertebrate prodynorphin precursor with vertebrate prodynorphin reveals a 28.8% sequence identity with rat, and 22% with the human and swine (Table 1).[43-45] In leech prodynorphin, α-neo-endorphin is found at position 67-76 and it exhibits a 100% sequence identity with the respective mammalian material. Dynorphin A-like material at 93-105 exhibits a 50% sequence identity and dynorphin B-like material at 106-117 exhibits a 76.6% sequence homology with the mammalian counterpart (Table 1). Although the α-neo-endorphin is identical to the one found in vertebrates, the dynorphins are slightly shorter. The amount of leu-enkephalin is similar to that found in vertebrates, i.e., three. Moreover, the C-terminus of leech prodynorphin is similar to that of vertebrates whereas the N-terminus is shorter.[44,45] This explains the difference in mass observed between the leech (14291 Da), rat (23386 Da),[43] swine (28616 Da)[46] and human prodynorphin (28385 Da),[47] suggesting that these additions occurred later in evolution.

As with the proenkephalin-derived peptides the leech prodynorphin-derived peptides are found in positions flanked by basic amino acids, indicating cleavage sites. Furthermore, the N-terminus of leech prodynorphin exhibits a 54.5% sequence homology with that of rat.[43-45] Also, leu-enk, α-neo-endorphin, dynorphin-A, dynorphin-B are present at the C-terminal side of the protein.

M. edulis also contains a prodynorphin molecule in its hemocytes.[44,45] *M. edulis* prodynorphin contains, a-neo-endorphin, dynorphin-A and dynorphin-B at the C-terminus, exhibiting 100%, 70.5%, and 85% sequence identity with the rat prodynorphin-derived counterparts, respectively.[44,45] The number of leu-enkephalins in this precursor is identical to that found in vertebrates. *M. edulis* prodynorphin is distinguished from that described earlier in leeches in that the N-terminus is longer. Additionally, the presence of an orphanin FQ-like peptide, exhibiting 50% sequence homology with that found in mammals, was demonstrated by sequence comparison.[45]

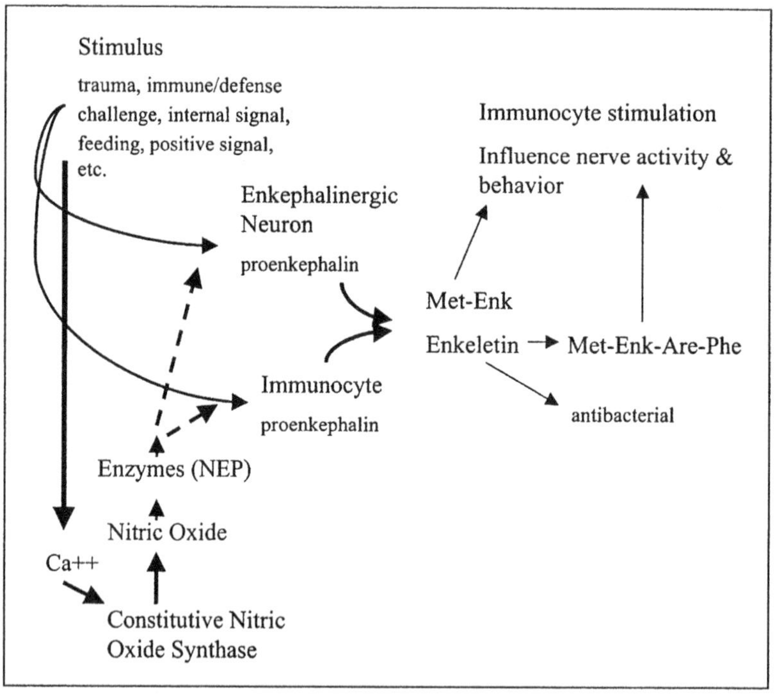

Fig. 2. Opioid peptide diversification of function. Various types of stimuli have the ability to initiate the enzymatic cleavage of constitutively expressed proenkephalin, liberating the antibacterial peptide enkelytin as well as the immunocyte stimulating opioid pentapeptide methionine-enkephalin.[31,33,2,34] Once secreted, simultaneously methionine-enkephalin stimulates immunocyte chemotaxis and activation while enkelytin exerts a bacteriocidal action. In time, enkelytin further is processed to methionine-enkephalin-Arg-Phe, which also exerts immunocyte stimulatory actions. Methionine-enkephalin also meditates neural processes that appear to be antinociceptive, dulling the sensation towards noxious stimuli (see text). In this regard, proenkephalin processing, the initial event, appears to be stimulated by an alteration of nitric oxide levels via constitutive nitric oxide synthase. We have shown that this increase in nitric oxide can stimulate neutral endopeptidase (NEP) activity via a calcium mediated process.[52,137] Thus, an alteration of constitutive nitric oxide levels, representing an innate immune protective response, may serve as the trigger for enhanced opioid processing by the metallo-endopeptidase NEP. This process also occurs during human surgery,[35] demonstrating the significance of this innate proactive immune and neural response.

Proopiomelanocortin (POMC)

Duvaux-Miret and colleagues[48] demonstrated the presence of β-endorphin and of a POMC-related gene in *Schistosoma mansoni*. Dot blots of cercarial genomic DNA, hybridized with two oligonucleotide probes complementary to highly conserved POMC sequences, showed a POMC-related gene in this parasite. Northern blot analysis of adult worm RNA indicated that this gene was actively transcribed and β-endorphin, ACTH, and α-melanocyte stimulating hormone (α-MSH) were detected in all developmental stages of the parasite by radioimmunoassay. Furthermore, *S. mansoni* secretes ACTH-like and β-endorphin-like peptides into its incubation medium.[49,50] This study constituted the first demonstration of a POMC-related gene transcript in an invertebrate.

The second report is by Salzet and colleagues[51] who sequenced a mammalian–like POMC, and six of its derived peptides, including ACTH and MSH, in the immune

Table 1. **Mytilus edulis *POMC and prodynorphin percent amino acid sequence identity with human and rat material***

Mytilus	Human	Rat	Mytilus	Human	Rat
POMC	35.3	40.26	Prodynorphin	35	55
ACTH	94.7	93.3	Nociceptin	25	35
γ-MSH	93	95	Orphanin FQ	50	50
α-MSH	100	100	β-Neo-endorphin	100	100
CLIP	95	95	α-Neo-endorphin	100	100
γ-LPH	5	5	Dynorphin A	70.5	70.5
β-endorphin	18	20	Dynorphin B	85.7	85.7
Met-enkephalin	100	100			

Sequence identity was first reported in the following reports.[45,52] ACTH, adrenocorticotropin; MSH, melanocyte-stimulating hormone; CLIP, corticotropin-like intermediate lobe peptide; LPH, lipotropin.

tissues of *T. tessulatum*. Of the six peptides, three showed high sequence similarity to their vertebrate counterparts, namely, met-enkephalin, α-MSH and ACTH (100, 84.6 and 70% respectively) whereas γ-MSH, β-endorphin and γ-lipotropic hormone (γ-LPH) exhibited only 45%, 20% and 10% sequence identity. No dibasic amino acid residues were found at the C-terminus of the γ- and β-MSH peptides. In contrast, the leech α-MSH was flanked at its C-terminus by the Gly-Arg-Lys amidation signal. ACTH and corticotropin-like intermediate lobe peptide (CLIP) were also C-terminally flanked by dibasic amino acid residues. The coding region of leech POMC was also reported by reverse transcription polymerase chain reaction (RT-PCR) using degenerated oligonucleotide primers.[51]

M. edulis hemocytes also contain a mammalian–like POMC (Fig. 1).[52] Of the six peptides found in this peptide precursor, Met-enkephalin, γ-MSH, α-MSH and ACTH exhibited 100%, 100%, 90% and 74% sequence identity, respectively (Table 1). The β-endorphin-like and γ-LPH-like molecules exhibit only 25% and 10% sequence identity. Dibasic amino acid residues are found at the C-terminus of MSH and ACTH, indicating cleavage sites. The α-MSH is flanked at the C-terminus by Gly-Arg-Lys, representing the amidation signal. ACTH and CLIP are also C-terminally flanked by dibasic amino acid residues. Of interest is the fact that the met-enkephalin present in the "β-endorphin-like peptide" is not flanked by dibasic amino acids, as is the case in mammals, suggesting that functionally it may be present. Taken together, the results from parasites and a free-living mollusk now conclusively demonstrate, without the presence of vertebrate "contamination", that POMC and many of the derived bioactive peptides, i.e., α-MSH, are present in invertebrates. Furthermore, in regard to their function, i.e., immune regulatory actions, they appear to be conserved as well.[36,51]

Opioid Receptors

Nervous Tissue

Opiate receptor sites present in the nervous tissue of *M. edulis* have properties very similar to those of opiate receptors in the mammalian nervous system.[1] In the first detailed account of opiate binding in invertebrates,[5,15,53] the binding of FK 33 824, an opiate agonist, and naloxone, an antagonist, were shown to be stereospecific, saturable, reversible, and of a high affinity. In contrast, specific binding to non-neural tissues, i.e., mantle, was negligible, suggesting that the opiate sites are restricted to nervous tissues. In addition, the binding of FK 33 824 was inhibited by sodium, a process that was reversed by manganese. The binding of naloxone was enhanced two-fold in the presence of a high sodium concentration and was relatively unaffected by manganese. These differential effects of ions on agonist and antagonist binding were similar to those observed in mammalian brain homogenates.

In a subsequent study,[15,53] the binding profiles obtained for various opiate ligands to membrane suspensions of pedal ganglia revealed the presence of both high- and low-affinity binding sites. FK 33 824 bound noncooperatively to a class of high-affinity sites (K_d = 1-3 nM) and cooperatively to a class of low-affinity sites (K_d = 6-11 nM). Hill analysis of the cooperative sites revealed Hill coefficients of n = 2.6-3.7, indicating markedly positive homotropic cooperativity. The total density of binding sites for all ligands was approximately 160 pmol/g of protein, whereby the high-affinity component comprised approximately 34% of the total. Kinetic analysis of the binding data obtained in *M. edulis* revealed values similar to those obtained for mammalian binding analysis. It also substantiated the cooperative nature of binding to the low-affinity site. The relative potencies of a series of opiates in displacing FK 33 824 enkephalin binding to membrane suspensions of pedal ganglia were very similar to those determined for rat brain homogenates.[1] The results also suggest that the high-affinity opiate binding sites, which mediate alteration in dopamine levels, are on presynaptic dopaminergic terminals.[54,55] The localization of opiate receptors on presynaptic nerve terminals has been documented for several areas of the mammalian nervous system (for review see ref. 13). Therefore, the opioid-dopamine interaction in *M. edulis* appears to be quite complex and analogous to the mechanisms existing in mammalian neural tissues.

Mu Opiate Receptor Transcripts

Supporting these biochemical and pharmacological studies involved with the demonstration of opiate receptors in *M. edulis*, we recently demonstrated, using molecular techniques, that this type of receptor subtype is present in the animal's neural tissues.[56] We examined the tissue with human specific primers designed to amplify a fragment of the coding region of the μ opiate receptor. RT-PCR and sequence analysis were used to detect μ-specific mRNA (Fig. 1). Using these μ-specific primers, a transcript for the _ receptor (441 bp) was amplified in RNA isolated from the ganglia. Sequence analysis of the PCR product demonstrates that this fragment exhibits 95% sequence identity with the human brain μ opioid receptor.[56] Furthermore, exposing pedal ganglia to both interleukin-1 -α and -β at 30 and 50 ng/ml for 24 hours resulted in a 47% and 60% increase in the band density, respectively.[56] Twenty-four hour treatment of pedal ganglia with morphine diminished the μ receptor transcript to almost zero.[56] In order to prevent any possible contamination with human tissue, we performed all invertebrate experiments in a designated area. In addition, extraction of total RNA from *Mytilus* tissues was done in a sterile environment. All of the tubes used were RNase and DNase-free and autoclaved. Furthermore, aerosol resistant tips (ART) were used to prevent the possibility of cross contamination. Reagents and enzymes used for theRT-PCR were used only for invertebrate studies. A negative reagent control was used with all RT-PCR reactions. The study suggests that the μ opiate receptor first evolved in invertebrates and was then conserved during evolution.

Table 2. Displacement of 3H-DAMA (1nM) by opioid ligands in Mytilus edulis immunocyte and pedal ganglia membrane suspensions

LIGAND	Immunocytes	Ganglia
Agonists		
δ-agonist		
DPDPE	9.1± 1.1	8.3± 0.9
Met-Enk	1.1± 0.1	0.9± 0.2
Met-Enk-Arg-Phe	1.3± 0.2	1.0± 0.1
DADLE	8.7±1.4	8.8±1.5
μ-agonist		
DAMGO	56±3.6	61± 4.2
Dihydromorphine	98 ± 13.1	106 ± 12.0
Morphine	145 ± 17.4	132 ± 15.7
κ-agonist		
Dynorphin 1-17	>22.7± 2.2	>24.6± 3.5
Antagonists		
Naltrexone	58 ± 9.8	74 ± 8.6
Naltrindole	1.1± 0.2	0.9± 0.2

Incubations with [Met]enkephalin contained phosphoramidon (100 mM) and bestatin (100 mM) to inhibit enzyme action (see 15).[40,53,63,133,134] The mean ± SD is noted. DAMA, D-Ala^2-Met^5-enkephalinamide; DPDPE, [D-Pen^2, D-Pen^5]-enkephalin; Met-ENK, met-enkephalin; DADLE, [D-Ala^2, Leu^5]-enkephalin; DAMGO, [D-Ala^2, MePhe^4, Gly(ol)^5]enkephalin)

Immune Cells

Delta Opioid-Type Receptors

In *M. edulis* immunocyte membrane suspensions a single high affinity ^{3}H-DAMA binding profile became evident upon Scatchard analysis.[40] This high affinity site has a K_d of 0.8 nM with a B_{max} of 4.0 pmol/g protein. The effects of pre-incubation of the immunocyte membranes with increasing concentrations of D-Ala^2, Leu^5, Cys^6-enkephalin (DALCE), a nonequilibrium delta opioid antagonist, on recovery of the delta binding sites of ^{3}H-D-Pen^2, D-Pen^5-enkephalin (^{3}H–DPDPE), ^{3}H-deltorphin I and ^{3}H-DAMA in *M. edulis* revealed two delta-type binding sites (Table 2).[40] DALCE pretreatment resulted in a marked, concentration-dependent decrease in recovery of ^{3}H-DPDPE-binding sites, whereas it only slightly modified the recovery of ^{3}H-deltorphin I-binding sites and ^{3}H-DAMA sites.[40]

The ability of a variety of other opioids to displace specifically bound ^{3}H-DAMA was investigated in another experiment (Table 2).[40] The opioid peptides were effective in the following decreasing order: deltorphin I = DAMA > Met-enkephalin> DADLE > DPDPE. By contrast, the μ and κ ligands D-Ala^2, MePhe^4, Gly(ol)^5-enkephalin (DAMGO) and dynorphin 1-17 were quite weak. Naltrindole was found to be more potent than naloxone in displacing ^{3}H-DAMA. Thus, as in human granulocytes, *M. edulis* immunocytes provide additional evidence that a special δ type of opioid receptor for immunoregulating Met-enkephalin exists on human granulocytes and their counterpart in invertebrates.[40] At

Table 3. Displacement of ^{3}H-dihydromorphine binding (IC_{50}, nmol/L) by opioid ligands in Mytilus edilus immunocytes and pedal ganglia membrane suspensions

LIGAND	Immunocytes	Ganglia
δ-agonist		
DPDPE	>1000	>1000
Met-Enk	>1000	>1000
Met-Enk-Arg-Phe	>1000	>1000
μ-agonist		
Endomorphin 1	>1000	>1000
Endomorphin 2	>1000	>1000
Nociceptin	>1000	>1000
DAMGO	>1000	>1000
Dihydromorphine	44 ± 8.1	42 ± 8.2
Morphine	42 ± 7.6	41 ± 8.0
κ-agonist		
Dynorphin 1-17	>1000	>1000
Antagonists		
Naltrexone	33 ± 5.3	36 ± 4.1
Naloxone	57 ± 8.3	63 ± 7.5

The mean + SD is noted. The displacement analysis data indicate the potency of various opioid extracts in displacing ^{3}H-dihydromorphine and provides specific information on different receptor populations. DPDPE, [D-Pen^2, D-Pen^5]-enkephalin; Met-ENK, met-enkephalin; DAMGO, [Tyr-D-Ala^2, Gly-N-Me-Phe^4, Gly(ol)^5)-enkephalin]. Combined from refs. 2,15,40,53,57,63,135.

that time, we decided that the special opioid receptor postulated to interact with Met-enkephalin in its modulatory function be tentatively classified as a subtype, δ_2, of the classical δ receptor.[40]

Mu Type Receptors

In 1991, first working with *M. edulis* immunocyte membranes, beside the δ_2 opioid receptor subtype, we found another type of receptor, designated μ_3.[57] The μ_3 receptor is distinguished from classical neuronal opioid receptor subtypes on the basis of pharmacological properties as revealed by radioligand competition and by functional studies, as well as by biochemical properties. Briefly, this receptor is opiate alkaloid selective and opioid peptide insensitive (Table 3).[36,57-61] The newly discovered opioid peptides endomorphin-1, -2 and orphanin FQ also do not bind to this opiate receptor subtype in vertebrate and invertebrate tissues (Table 3).[62,63] 6-Glucuronide but not the 3-glucuronide metabolite of morphine, binds to the μ_3 receptor in invertebrates (unpublished). Following its discovery in invertebrate tissues we found it also to be present on human monocytes[57] and later, on many other types of cells, i.e., human endothelia.[60]

As with classical opioid receptors,[64] μ_3 is linked to trimeric G proteins that, in turn, have the capability to modulate Ca^{++} and K^+ channels, adenylyl cyclase, and probably other signal transduction systems.[65] Recently, we have demonstrated that this opiate receptor subtype is coupled to intracellular calcium transients,[66] supporting a classical μ signaling pattern.

Opioid Processing

We surmise that the opioid precursor proteins are sequestered in the cell and only processed into their smaller active peptides, such as Met-enkephalin, when required (see ref. 2). The functions of these molecules in invertebrates can be deduced from the processing of the precursor molecules in various tissues. Since precursor processing involves enzymes, the presence of specific enzymes becomes important. An examination of the literature reveals the presence of many types of enzymes in both vertebrates and invertebrates, some of which are important in processing neuropeptides, e.g., neutral endopeptidase (NEP; Fig. 2) and angiotensin converting enzyme (ACE) (see refs. 20,21,67-70).

The enzyme NEP appears to be quite important. For example, it may not only be responsible for cleaving the precursor proenkephalin or POMC/ACTH, but the active processed peptides as well, i.e., Met-enkephalin and MSH, respectively.[49,71-73] This represents a multidimensional process that requires less DNA "messaging" since the same enzyme performs these tasks. Furthermore, in some cases, the actual inactive products may act as competitive inhibitors to further limit the activity of the prime enzyme, adding another degree of microenvironmental control. This has been noted in our laboratory by NEP processing of Met-enkephalin-Arg-Phe.[72]

The significance of these observations is illustrated by the actions of aprotinin, a serine protease inhibitor.[74] The use of this compound can and does diminish the diffuse inflammatory response associated with surgery,[75,76] demonstrating the significance of processing enzymes. In patients ready to undergo major heart surgery, we found that just before surgery, plasma ACTH levels dropped below the level of detection (see ref. 77), indicating the activation of the processing enzymes (see ref. 73). In this regard, it is widely known that various immune and neural-type signaling molecules can up-regulate enzymes such as NEP.[20,36,78] This response is biphasic. First, a mechanism for enhanced neuropeptide precursor processing occurs, followed by enhanced processed peptide degradation due to a further increase in enzyme levels, i.e., immunocyte recruitment. Clearly, with this scenario, cascading immune responses can be better understood. In this regard, it is important to realize that invertebrate immune/defense systems have been utilizing these processes for over 500 million years (see ref. 31).

There is a growing body of evidence demonstrating that morphine influences ACTH processing in vertebrates and invertebrates (see ref. 36). This is especially important since it is a naturally occurring signal molecule found in human plasma and invertebrate hemolymph.[57,79-81] In this regard, exogenously applied morphine specifically can increase the release of ACTH in rat hypothalamus (see ref. 36).[82] Recent work from our laboratory demonstrates that NO controls neurohormonal release from median eminence neuroendocrine nerve terminals in the rat.[83] The stimulation of this release from median eminence fragments, including vascular tissues, occurs by μ_3 receptor activation by morphine. Furthermore, morphine, by the NO dependent process, influences neurohormonal release from median eminence nerve terminals within 10 minutes releasing corticotropin which can then account for the action of morphine noted earlier.

In invertebrates, morphine has been shown to increase ACTH hemolymph levels,[52] probably by increasing the processing of POMC or the release of this peptide from immunocytes (Fig. 2). Morphine, in a dose-dependent manner, and by way of NO, increased leech processing of POMC as noted by higher hemolymph levels of α-MSH and ACTH.[51] In *M. edulis* we also demonstrate that morphine stimulates the processing of ACTH (1-39) to MSH (1-13) by NEP as determined by phosphoramidon inhibition. The ability of morphine to enhance enzyme levels has also been noted in other studies using mammalian and human tissues (see refs. 36,84). The mechanism for this morphine action, based on these reports, is by increasing the processing of the precursor or stimulating the release of the precursor, or both.

The significance and specificity of opiate molecules in these studies are enhanced by the observation that lipopolysaccharide (LPS; endotoxin derived from the cell wall of gram-negative bacteria) stimulation results in increased levels of ACTH (1-24) in the hemolymph, indicating that other enzymatic processes can occur by way of different signaling molecules. Furthermore, ACTH (1-24) processing occurred by an enzymatic process independent of NEP, i.e., renin-type enzyme.[34,52] Taken together, as in mammals, differential processing of ACTH occurs in invertebrates. Additionally, invertebrate immunocytes are capable of displaying different responses to ACTH fragments, including those of *M. edulis*,[85] further supporting the differential processing pattern and its potential significance as a meaningful event.

Immune Opioid Peptide Actions

Immune System

Chemotactic effects of endogenous opioid peptides on human polymorphonuclear leukocytes, monocytes, and lymphocytes have been demonstrated.[86-89] Moreover, Stefano and colleagues[38,39,90] have demonstrated that opioids, by stereoselect mechanisms, are involved in invertebrate autoimmunoregulatory processes. Interestingly, a subpopulation of granulocytes and immunocytes from *M. edulis* and *Leucophaea maderae* has the ability to respond to low opioid concentrations by adhering and clumping.[38,39] The adherence-promoting role of DAMA and its blockage by naloxone, in a dose response manner, were clearly evident. By contrast, exogenous met-enkephalin at the same low concentration of DAMA, did not increase cellular adherence above control levels, due to the presence of proteolytic enzymes in the hemolymph.[20] Subsequent studies demonstrated that indeed neutral endopeptidase 24.11 (CD10, "enkephalinase") was present on both human and invertebrate immunocytes[20] where it serves to modulate neuropeptide activation of the respective cells.[72,78] Cytokine neural stimulation in *M. edulis* results in up-regulating of neutral endopeptidase activity which in turn down-regulates the cells responsiveness to various neuropeptide substrates of this enzyme.[72,78,91] Clearly, given this type of complex regulation the importance of an autoimmunoregulation role for morphine is enhanced.

Severing of nerves can elicit an immune response in *M. edulis*.[38] Nerve severance evoked a cellular immune response, as judged by the directional migration of yellow-fluorescent immunocytes to the lesioned area. The concentration of these cells accumulating and adhering to the lesioned tissue gradually increased, a response presumed to be due to a concentration gradient of antigenic or recognition factors. An injection of DAMA, placed in the vicinity of a severed nerve, showed that, after a period of two hours, the concentration gradient established by the injected material had taken precedence over that provided by putative endogenous antigenic messengers dispatched at the site of lesion. A possible explanation for this differential response is a critical difference between the concentrations of endogenous and injected ligands competing for opioid receptors. Subsequently, it was demonstrated with in vitro tests that the stimulation of locomotory behavior of invertebrate immunocytes by opioids is accompanied by distinctive conformational changes. Such changes (flattening, increase in surface area), resembling those reported in mammals (see above), also occur in unstimulated preparations, but at a lower frequency. The in vivo tests in *M. edulis* referred to above, indicate that the administration of exogenous opioid material may elicit a directed movement of immunocytes. Similarly, cellular stimulation by various opioid drugs in slide tests reveals directed, as well as random, locomotion. While unstimulated immunocytes showed some random movements, clumping occurred only in the presence of opioids. This may be taken as evidence for the occurrence of chemotactic, as well as chemokinetic, activities of opioids. However, the participation of a second signal molecule, giving direction to randomly migrating cells[92] cannot be ruled out. Evidence for the presence of opioid receptors in the immunocytes of *M. edulis* and *L. maderae* studied, was obtained by determining the effects of naloxone on the cellular activities under consideration. Naloxone injections into the area of nerve severance of *M. edulis* noted above, counteracted the cellular immune reaction observed in the absence of this drug[38,39] Thus, immunocytes containing opioid peptides have the capability of responding to them as well.

Monokines

Previous reports have demonstrated the presence, in starfish and tunicates, of factors with interleukin (IL)-1-like effects.[93] Since *M. edulis* immunocytes most resemble monocyte/macrophages, the effects of the human monokines, tumor necrosis factor (TNF) and IL-1, were determined on these cells.[92] It was demonstrated that *M. edulis* immunocytes respond to these substances, both in vitro and in vivo, in a fashion similar to human granulocytes. TNF in a dose dependent fashion, increases the relative reflectance of the cells and also causes the cells to flatten as indicated by the measured increase in their area and their perimeter. Recombinant human IL-1 initiates responses similar to TNF in *M. edulis* immunocytes. In addition, it appears that the immunocytes respond to IL-1, at least in part, through TNF production. Finally, immunoreactive TNF and IL-1 were detected in *M. edulis* hemolymph. Szucs and colleagues[94] demonstrate that cytokines can effect invertebrate neurons. In another report in this issue Paeman and coworkers[91] find immunoreactive IL-1 in glial cells present in the ganglia of *M. edulis* and *Neries sp* (a marine worm). This finding in *M. edulis* corroborates a previous study demonstrating that DAMA can stimulate the secretion of an IL-1-like molecule from *M. edulis* pedal ganglia.[95] Additionally, in regard to the presence of cytokine-like molecules in invertebrate ganglia, the specific anatomical localization of these cytokine-like molecules in the posterior central portion of *M. edulis* pedal ganglia provides the foundation for a neuroimmune connection reported in another work, strongly suggesting this interaction following electrical stress.[96] Thus, evidence is accumulating which supports not only the concept of neuroimmunology developing in invertebrates, but that the invertebrate immune system shares autoimmunoregulatory characteristics with mammals. The above results provide information on the evolutionary history of basic biological phenomena and, on occasion, point the way to important insights applicable also to higher organisms, including mammals.

Morphine

The demonstration of endogenous opiates, i.e., morphine, codeine, in various vertebrate and invertebrate tissues, including the nervous system (e.g., see refs. 57,79,97-107), is quite important for establishing the significance of the μ_3 opiate receptor subtype as well as the signaling status of this endogenous chemical messenger.[105,106] Besides these biochemical studies, immunocytochemical localization of a morphine-like material was reported in neural and immune tissues,[80,108,109] as well as in other invertebrate tissues.[57,104-106]

In a recent report[104], we demonstrate the presence of a morphine-like compound by biochemical and immunocytochemical methods in a freshwater snail (*Planorbarious corneus*). Using a high pressure liquid chromatography (HPLC) coupled to an electrochemical detector, the pedal ganglia morphine-like level was determined (6.20 ± 2.0 pmol/g). This material was also found in immune and muscular tissues, but not in the animal's hepatopancreas. In animals which were traumatized by the cutting of their foot sole, the ganglionic morphine levels rose significantly after 24 hours (43.7 + 5.2 pmol/g, $P < 0.005$) and 48 hours (19.3 + 4.6 pmol/g, $P < 0.05$). Simultaneously, the ganglionic morphine-immunoreactivity (morphine-IR) increased in both intensity and the number of structures responding positively, i.e., neurons and fiber varicosities. The morphine-IR and biochemical levels also increased in immune cells and peripheral nerves. In another mollusk, *Mytilus galloprovincialis*, the same pattern of enhanced morphine-IR was found after trauma. Taken together, the study demonstrated the presence of a morphine-like compound in neural and immune tissues that increased after trauma.

Recently, we have found morphine and morphine-6-glucoronide (M6G), a morphine metabolite, in the pedal ganglia of *Modiolus deminissus*, another marine bivalve, at a level of 2.41 ng/ganglia and 0.95 ng/ganglia, respectively.[105] These opiate alkaloids are normally found at low concentrations in invertebrate and vertebrate tissues, including neural. Given this problem, we also described a new opiate extraction protocol as well as a HPLC purification procedure that can separate and quantify morphine and its derivatives at sub-nanogram concentrations. Furthermore, both morphine and M6G were identified in this mollusk's pedal ganglia by mass spectrometry analysis.[105] Additionally, codeine has also been found in *Mytilus edulis*.[57,79]

Using a gradient of acetonitrile, M6G, morphine and morphine 6-acetate (MA, not naturally occurring) eluted at 9.6%, 19.4% and 91.1% of B buffer, respectively from *M. edulis* pedal ganglia extracts (i.e., 4.8%, 9.7% and 45.6% of acetonitrile, respectively; Fig. 3A). The M6G, morphine and MA standard curves exhibit a linearity curve cooperation of 0.948, 0.998 and 0.997, respectively (Fig. 3B). Furthermore, for the low concentration of morphine a cooperation of 0.983 was observed (35–280 pg) (Fig. 3A inset). The lowest amount of M6G detected is 80 pg with a ratio signal/noise (s/n) of 2.15 and the limit of quantification at 125 ng (ratio s/n=3.72). The lowest amount of morphine detected is 20 pg with a s/n ratio of 1.89 and the quantification limit of 35 pg (s/n=3.9). Finally, the MA detection limit is 125 pg (s/n=3.2) and the quantification limit is 250 pg with a s/n of 4.13. Additionally, this extraction method gives a recovery of 84.3% (+/- 9.1%, n=4), 86.2% (+/- 6.2%, n=4), and 78.4% (+/- 10.9%, n=4) for M6G, morphine and MA respectively, in comparison between the internal and external standards.

Ten pedal ganglia from *M. edulis* were examined for their endogenous opiate alkaloid levels. The resulting sample, following extraction, was then purified by HPLC as noted earlier. The chromatogram obtained (Fig. 3C) had specific peaks corresponding to M6G and morphine. These peaks were collected, coded and subjected to gas-chropatography mass spectrometry (GC/MS) and commercial analysis. The GC/MS (Fig. 3D) and independent analysis confirmed the presence of M6G as well as morphine in peaks 1 and 2 (Fig. 3C), respectively. Quantification of morphine and morphine-derivatives using the software Chromatogram Report demonstrated an amount of M6G and morphine at 2.67 + 0.44 ng/ganglia and 0.98 + 0.14 ng/ganglia, respectively. Quantification of MA was not performed because this compound is derived from heroin and is not naturally present in living organisms.

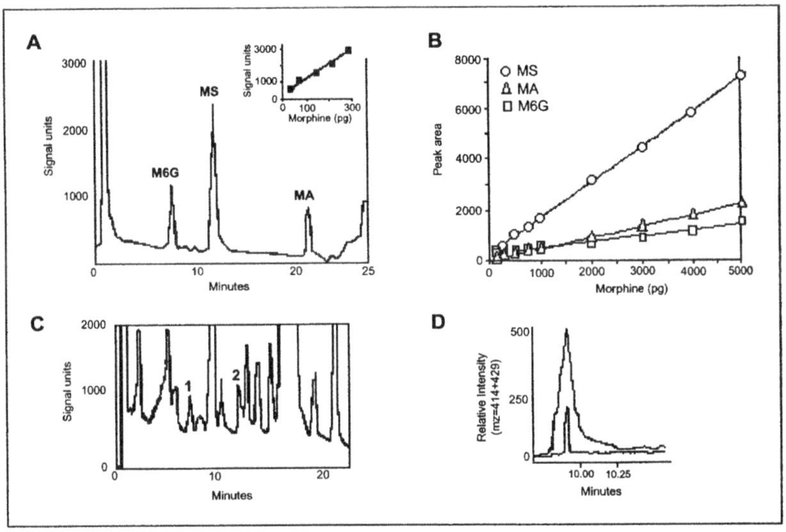

Fig. 3. Biochemical and physical analysis of opiate alkaloids from Mytilus neural tissues. A. Chromatogram of 5 ng of M6G (morphine-6-glucoronide), MS (morphine) and MA (morphine 6-acetate); inset, Low concentrations of MS is also linear and a function of the area determined for in the range of 35 to 280 pg. B. Concentration of M6G, MS and MA is a function of the area of the peak determined for the concentration range of 125 to 5000 pg. C. Separation of the material extracted form *Mytilus edulis* pedal ganglia. Peak 1 and 2 correspond to M6G and MS, respectively. These peaks were collected and submitted for mass spectrometry analysis. D. Gas-chromatography mass spectrometry (GC/MS) spectrum of morphine standard (400 pg, larger curve) and the morphine material collected during high pressure liquid chromatography (HPLC) analysis of *Mytilus* pedal ganglia. Extraction experiments using internal or external morphine standards were performed in a different room to avoid morphine contamination of the biological samples tested. Single use siliconized tubes were used to prevent the loss of morphine. Tissues were extensively washed with PBS buffer (3 times, 1min) to avoid exogenous morphine contamination. Blanks and nonmorphine containing tissues, i.e., mantle, were run via HPLC between runs to determine and remove any residual morphine. Furthermore, the fraction of blank chromatography corresponding to the elution time of morphine (flat line) was check by mass spectrometry analysis, confirming that no morphine remained. GC/MS confirmed the identity of morphine.[105,107,136] All techniques employed are as described elsewhere in detail.[105,107,136] Mass spectrum indicated major ions (base peak depended on instrument tune conditions) at 429 (M+) and 414 (M-CH_3+). Analyses of samples were carried out using a selected ion storage method, in which mass windows (+/- 2 amu) around ions 429 and 414 were collected. Morphine identity was confirmed by the retention time, peak shape, and comparison of pedal ganglia derived morphine to authentic morphine standards injected. Furthermore, collected HPLC-fractions were coded and independently analyzed to confirm the presence of MS and M6G (Anaspec Incorporated, San Jose, California).

Thus, morphine can be found in plants, invertebrates and vertebrates, including human tissues,[106] suggesting that it may have been in a common ancestor before animals and plants split in evolution.

Besides the presence of morphine in free-living invertebrates, it has also been found in parasitic worms. It has been identified as a morphine-like molecule in *S. mansoni* by way of radioimmunoassay.[110] The parasitic worm *Ascaris suum* contains the opiate alkaloid morphine as determined by HPLC coupled to electrochemical detection and by GC-MS.[107] The level of this material is 1168 ± 278 ng/g worm wet weight. Furthermore, *A. suum* maintained for five days contained a significant amount of morphine, as did their medium, demonstrating their ability to synthesize and to secrete the opiate alkaloid. To determine if the morphine was active, we exposed human monocytes to the material and they immediately released nitric oxide in a naloxone-reversible manner. The anatomic distribution of morphine-IR reveals that the material is in the subcuticle layers and in the animals' nerve chords.[107] Furthermore, as determined by RT-PCR, *A. suum* does not express the transcript of the neuronal μ receptor.[107] Failure to demonstrate the expression of this opioid receptor, as well as the morphine-like tissue localization in *A. suum*, suggests that the endogenous morphine is intended for secretion into the microenvironment. Taken together, these data indicate that morphine is used to diminish the capacity of the host response to the presence of the parasite, including dampening the alerting neural and immune processes. Another recent report demonstrates opiate alkaloid processes in a leech,[111] confirming our recent study.

Nitric Oxide (NO)

There has been in the literature an association of NO with morphine actions. Peripheral morphine antinociception (analgesia) involves NO-stimulated increases in intracellular cyclic guanosine 5'-monophosphate (cGMP).[112] Nitric oxide has been associated with nociception[113] as well as tolerance and dependence.[114] In addition, the morphine-induced suppression of splenic lymphocyte proliferation has been shown to involve NO.[115] Morphine and NO have been linked in gastrointestinal regulation.[116] Furthermore, morphine, not opioid peptides, stimulates constitutive NO release in macrophages, granulocytes, various types of human and rat endothelial cells, invertebrate neurons and immunocytes and in rat median eminence fragments, all in a naloxone antagonizable manner.[36,62,81,83,117-122] These data suggest that the $\_{3}$ receptor is coupled to constitutive NO release in these cells.

Opiate Immune Actions

Furthermore, morphine's actions in these diverse tissues complements what is known about NO mediating immune and vascular functions, namely that it can down regulate them from an excitatory state or prevent the excitatory state from occurring.[36,103,120,122-125] Additional information on opiate alkaloid signaling substances can be summarized as follows: Injection of vertebrate animals with morphine results in deficient macrophage function[126] and an alteration of T-cell activity.[127] Morphine also antagonizes IL-1α- or TNF-α-induced chemotaxis in human granulocytes and monocytes.[128,129]

Morphine down-regulates invertebrate immunocytes, causing active motile amoeboid cells to become round and immobile.[57,79] It also diminishes invertebrate microglial activation and egress from ganglia maintained in vitro.[41] Taken together, morphine inhibits invertebrate immunocytes. However, it must be emphasized that our observations in this regard are continuous. This is important, because following the inhibitory action, the cells rebound into excitation.

Noxious Stimuli

It comes as no surprise that an amoeba can sense an aversive environment and move away from heat, acid, etc. Clearly, all living cells probably express this or a modified form of this innate protective survival ability (adaptation responses), as do organisms. This is clearly a response that all successful organisms must have. Specifically, in *M. edulis*,[130] touching the animals incurrent siphon causes its momentary withdrawal whereas cutting it not only causes it to undergo writhing motions, but the animal closes its valves for a prolonged time. This would tend to indicate the organism has the ability to differentiate the degree of stimulus strength, i.e., noxious (painful) stimulus. In another invertebrate, Kavaliers and colleagues[131] found that snails respond to a hot plate test by lifting their anterior body region. Additionally morphine delayed this response in a naloxone antagonizable manner, suggesting that opiates can modulate this noxious experience much the same as in mammals. Thus, it would appear that invertebrates can sense and avoid noxious stimuli and that; at least in some examples of this phenomenon, opiate signaling appears to be involved. As to whether these animals can experience a mammalian-type of pain, remains unclear.

In this regard, stress can be viewed as a threatening or harmful experience that may be cognitive or noncognitive. Organisms can be stressed by restraint, exposure to cold or heat, electric shock, etc. Such stimuli create tension and call into action defense mechanisms aiming to overcome them. In mammals, mechanisms recognized to be of primary importance in responding to stress and effecting the involvement of the immune defense system are the sympathetic-adrenomedullary system and the hypothalamic-hypophyseal-adrenocortical axis. In addition to hormones, other signal molecules participate in this process. The degrees of cellular immune response evoked can be taken as a measure of the severity of the disturbance and the organism's capacity to cope with it. In *M. edulis*, we noted earlier, the organism could respond appropriately to electrical stress. Furthermore, given the presence of POMC and ACTH it comes as no surprise to find these molecules and opioid peptides involved with this response.[96,132] Thus, the noncognitive substrate of the mammalian stress response may also have evolved earlier than previously thought.

In this light, we also can consider internal and external stimuli that upset an organism's homeostasis as being normal as long as it falls into the organism range of dynamic response. These environments are never stable for a long period of time and are designed, via millions of years of evolution, to absorb these changes by way of adaptation responses. In many instances these adaptation responses have been associated with the term stress, implying a negative phenomenon. Thus, the term stress may be inappropriate for many of these normal adaptational responses. In part, this may help explain the reluctance to accept these processes in simpler animals.

Conclusions

In summary, opioid and opiate immune processes appear to have had an earlier start in evolution than formerly realized. Additionally, given the presence of the components of these signaling systems, i.e., receptors, stereospecificity may be the actual "glue" maintaining these systems during evolution. The high opiate alkaloid selectivity of the μ_3 opiate receptor subtype reinforces a role and the presence of endogenous morphine. In regard to their immunomodulation, it appears that the opiate alkaloids inhibit, whereas opioid peptides tend to stimulate invertebrate and vertebrate immune cells, including proinflammatory cytokine production, operating in an antagonistic manner unlike their analogous analgesic actions. Parasites appear to be using these same host signal molecules to escape host immunosurveillance, further highlighting the activity of these molecules in the host as diminishing immune and neural actions. In regard to adaptational responses, again the same molecules that are involved in this response in mammals appear to be present in invertebrates and functioning in a similar capacity. Thus, we are left with the conclusion that many of these signal molecules and their functions had their origins in "simple" animals.

Acknowledgements

This work was supported by the following grants: NIMH COR 17138; NIDA 09010; NIMH 47392 and the Research Foundation and Central Administration of the State University of New York. We wish to acknowledge the technical assistance of Mr. Mazen Madhoun, Ms. Ireen Khan, Ms. Hiwet Mekonnen (NIMH-COR Fellows) and Ms. Danielle Benz. Dr. Wei Zhu is a Fellow of the Mind-Body Medical Institute of the Beth Israel Deaconess Medical Center.

References

1. Stefano GB. Comparative aspects of opioid-dopamine interaction. Cell Mol Neurobiol 1982; 2:167-178.
2. Stefano GB, Salzet M. Invertebrate opioid precursors: evolutionary conservation and the significance of enzymatic processing. Int Rev Cytol 1999; 187:261-286.
3. Stefano GB. Endocannabinoid immune and vascular signaling. Acta Pharmacol Sinica 2000; 21:1071-1081.
4. Stefano GB, Stefano JM. Cannabinoids: New antihypertensive agents? Curr Opin Clinical Exper Res 2000; 2:205-223.
5. Stefano GB. Opiates and neuroactive pentapeptides: Binding characteristics and interactions with dopamine stimulated adenylate cyclase in the pedal ganglia of *Mytilus edulis*. In: Rozsa KS, ed. Neurotransmitters in Invertebrates. London: Pergamon Press, 1980:423-453.
6. Stefano GB. Conformational matching: A determining force in maintaining signal molecules. In: Stefano GB, ed. Comparative Opioid and Related Neuropeptides Mechanisms. Boca Raton: CRC Press Inc., 1986:271-277.
7. Stefano GB. Neurobiology of Mytilus edulis. Manchester: Univ of Manchester, 1990.
8. Frontali N, Gainer H. Peptides in invertebrate nervous system. In: Gainer H, ed. Peptides in Neurobiology. New York: Plenum Press, 1977:259-271.
9. Haynes LW. Peptide neuroregulation in invertebrates. Prog Neurobiol 1980; 15:205-223.
10. Scharrer B. The neurosecretory neuron in neuroendocrine regulatory mechanisms. Am Zool 1967; 7:161-168.
11. Scharrer B. Peptidergic neuron: Facts and trends. Gen Comp Endocrinol 1978; 34:50-62.
12. Audigier Y, Deprat AM, Cros J. Comparative study of opiate and enkephalin receptors on lower vertebrates. Comp Biochem Physiol 1980; 67:191-194.
13. Leung MK, Stefano GB. Comparative neurobiology of opioids in invertebrates with special attention to senescent alterations. Prog Neurobiol 1987; 28:131-159.
14. LeRoith D, Trudgill DL, Roth J et al. Corticotropin and β-endorphin-like materials are native to unicellular organisms. Proc Natl Acad Sci USA 1982; 79:2086-2090.
15. Kream RM, Zukin RS, Stefano GB. Demonstration of two classes of opiate binding sites in the nervous tissue of the marine mollusc *Mytilus edulis*. Positive homotropic cooperativity of lower affinity binding sites. J Biol Chem 1980; 255:9218-9224.
16. Pert CB, Taylor D. Type 1 and type 2 opiate receptors: a subclassification scheme based upon GTP's differential effects on binding. In: Way EL, ed. Endogenous and Exogenous Opiate Agonists and Antagonists. New York: Pergamon Press, 1980:87-94.
17. Edley SM, Hall L, Herkenham M et al. Evolution of striated opiate receptors. Brain Res 1982; 249:184-188.
18. Stefano GB, Scharrer B, Leung MK. Neurobiology of opioids in *Leucophaea maderae*. In: Cockroaches as models for neurobiology: Applications in Biomedical Research. Boca Raton: CRC Press Inc., 1989:85-102.
19. Salzet M, Bulet P, Verger-Bocquet M et al. Isolation and structural characterization of enkephalin-related peptides in the brain of the Rhynchobdellid leech *Theromyzon tessulatum*. FEBS Lett 1995; 357:187-191.
20. Shipp MA, Stefano GB, D'Adamio L et al. Downregulation of enkephalin-mediated inflammatory response by CD10/neutral endopeptidase 24.11. Nature 1990; 347(6291):394-396.
21. Turner A, Leung MK, Stefano GB. Peptidases of significance in neuroimmunoregulation. In: Scharrer B, Smith EM, Stefano GB, eds. Neuropeptides in neuroimmunology. Springer-Verlag, 1994:152-169.
22. Schoofs L, Veelaert D, Vanden Broeck J et al. Peptides in the locusts, *Locusta migratoria* and *Schistocerca gregaria*. Peptides 1997; 18(1):145-156.

23. Gade G, Hoffmann KH, Spring JH. Hormonal regulation in insects: facts, gaps, and future directions. Physiol Rev 1997; 77(4):963-1032.
24. Zhu W, Vandingenen A, Huybrechets R et al. In vitro degradation of the Neb-Trypsin Modulating Oostatic Factor (Neb-TMOF) in gut luminal content and hemolymph of the grey fleshfly, *Neobellieria bullata.* Insect Biochem Molec Biol 2001; 31(1):87-95.
25. Leung MK, Stefano GB. Isolation and identification of enkephalin in pedal ganglia of *Mytilus edulis* (mollusca). Proc Natl Acad Sci USA 1984; 81:955-958.
26. Stefano GB, Leung MK. Presence of met-enkephalin-Arg-Phe in molluscan neural tissues. Brain Res 1984; 298:362-365.
27. Luschen W, Buck F, Willig A et al. Isolation, sequence analysis, and physiological properties of enkephalins in the nervous tissue of the shore crab, *Carcinus maenas* L. Proc Natl Acad Sci USA 1991; 88:8671-8675.
28. Rothe H, Luschen W, Asken A et al. Purified crustacean enkephalins inhibits release of hyperglycemic hormone in the crab, *Carcinus maenas* L. Comp Biochem Physiol 1991; 99C:57-62.
29. Laurent V, Salzet M. Metabolism of enkephalins in head membranes of the leech *Theromyzon tessulatum* by peptidases. Reg Peptides 1996; 65:123-131.
30. Ewandinger NM, Ridgway RL, Syed NI et al. Identification and localization of a (Met5)-enkephalin-like peptide in the mollusc, *Lymnaea stagnalis.* Brain Res 1996; 737:1-15.
31. Salzet M, Stefano GB. Invertebrate proenkephalin: Delta opioid binding sites in leech ganglia and immunocytes. Brain Res 1997; 768(224):232.
32. Goumon Y, Strub JM, Moniatte M et al. The C-terminal biophosphorylated proenkephalin-A-(209-237)-peptide from adrenal medullary chromaffin granules possesses antibacterial activity. Eur J Biochem 1996; 235(3):516-525.
33. Stefano GB, Salzet B, Fricchione GL. Enkelytin and opioid peptide association in invertebrates and vertebrates: immune activation and pain. Immunol Today 1998; 19(6):265-268.
34. Tasiemski A, Verger-Bocquet M, Cadet M et al. Proenkephalin A-derived peptides in invertebrate innate immune processes. Mol Brain Res 2000; 76(2):237-252.
35. Tasiemski A, Salzet M, Benson H et al. The presence of opioid and antibacterial peptides in human plasma during coronary artery bypass surgery. J Neuroimmunol 2000; 109:228-235.
36. Stefano GB, Scharrer B, Smith EM et al. Opioid and opiate immunoregulatory processes. Crit Rev in Immunol 1996; 16(2):109-144.
37. Zhong F, Li XY, Yang S et al. Methionine-enkephalin stimulates interleukin-6 mRNA expression: Human plasma levels in coronary artery bypass grafting. Int J Cardiol 1998; 64(1):S53-S59.
38. Stefano GB, Cadet P, Scharrer B. Stimulatory effects of opioid neuropeptides on locomotory activity and conformational changes in invertebrate and human immunocytes: Evidence for a subtype of delta receptor. Proc Natl Acad Sci USA 1989; 86:6307-6311.
39. Stefano GB, Leung MK, Zhao X et al. Evidence for the involvement of opioid neuropeptides in the adherence and migration of immunocompetent invertebrate hemocytes. Proc Natl Acad Sci USA 1989; 86:626-630.
40. Stefano GB, Melchiorri P, Negri L et al. (D-Ala2)-Deltorphin I binding and pharmacological evidence for a special subtype of delta opioid receptor on human and invertebrate immune cells. Proc Natl Acad Sci USA 1992; 89(19):9316-9320.
41. Sonetti D, Ottaviani E, Bianchi F et al. Microglia in invertebrate ganglia. Proc Natl Acad Sci USA 1994; 91:9180-9184.
42. Salzet M, Verger-Bocquet M, Bulet P et al. Purification, sequence analysis and cellular localization of a prodynorphin-derived peptide related to the α-neo-endorphin in the rhynchobdellid leech *Theromyzon tessulatum.* J Biol Chem 1996; 271:13191-13196.
43. Civelli O, Douglass J, Goldstein A et al. Sequence and expression of the rat prodynorphin gene. Proc Natl Acad Sci USA 1985; 82(12):4291-4295.
44. Salzet M, Stefano GB. Prodynorphin in invertebrates. Mol Brain Res 1997; 52:46-52.
45. Stefano GB, Salzet-Raveillon B, Salzet M. *Mytilus edulis* hemolymph contain prodynorphin. Immunol Lett 1998; 63:33-39.
46. Horikawa S, Takai T, Toyosata M et al. Isolation and structural organization of the human preproenkephalin B gene. Nature 1983; 306(5943):611-614.
47. Kakidani H, Furutani Y, Takahashi H et al. Cloning and sequence analysis of cDNA for porcine beta-neo-endorphin/dynorphin precursor. Nature 1982; 298(5871):245-249.
48. Duvaux-Miret O, Dissous C, Guatron JP et al. The helminth *Schistosoma mansoni* expresses a peptide similar to human beta-endorphin and possesses a POMC-related gene. New Biol 1990; 2:93-99.

49. Duvaux-Miret O, Stefano GB, Smith EM et al. Immunosuppression in the definitive and intermediate hosts of the human parasite *Schistosoma mansoni* by release of immunoactive neuropeptides. Proc Natl Acad Sci USA 1992; 89:778-781.
50. Duvaux-Miret O, Stefano GB, Smith EM et al. Proopiomelanocortin-derived peptides as tools of immune evasion for the human trematode *Schistosoma mansoni*. Acta Biol Hungari 1992; 43:281-286.
51. Salzet M, Cocquerelle C, Verger-Bocquet M et al. Leech immunocytes contain proopiomelanocortin: nitric oxide mediates hemolymph POMC processing. J Immunol 1997; 159(11):5400-5411.
52. Stefano GB, Salzet-Raveillon B, Salzet M. *Mytilus edulis* hemocytes contains pro-opiomelanocortin: LPS and morphine stimulate differential processing. Mol Brain Res 1998; 63:340-350.
53. Stefano GB, Kream RM, Zukin RS. Demonstration of stereospecific opiate binding in the nervous tissue of the marine mollusc *Mytilus edulis*. Brain Res 1980; 181:445-450.
54. Stefano GB, Hall B, Makman MH et al. Opioid inhibition of dopamine release from nervous tissue of *Mytilus edulis* and *Octopus bimaculatus*. Science 1981; 213(4510):928-930.
55. Stefano GB, Zukin RS, Kream RM. Evidence for the presynaptic localization of a high affinity opiate binding site on dopamine neurons in the pedal ganglia of *Mytilus edulis*. J Pharmacol Exp Ther 1982; 222:759-764.
56. Cadet P, Stefano GB. *Mytilus edulis* pedal ganglia express μ opiate receptor transcripts exhibiting high sequence identity with human neuronal $\mu 1$. Mol Brain Res 1999; 74:242-246.
57. Stefano GB, Digenis A, Spector S et al. Opiatelike substances in an invertebrate, a novel opiate receptor on invertebrate and human immunocytes, and a role in immunosuppression. Proc Natl Acad Sci USA 1993; 90:11099-11103.
58. Cruciani RA, Dvorkin B, Klinger HP et al. Presence in neuroblastoma cells of a μ_3 receptor with selectivity for opiate alkaloids but without affinity for opioid peptides. Brain Res 1994; 667:229-237.
59. Dobrenis K, Makman MH, Stefano GB. Occurrence of the opiate alkaloid-selective μ_3 receptor in mammalian microglia, astrocytes and kupffer cells. Brain Res 1995; 686:239-248.
60. Stefano GB, Hartman A, Bilfinger TV et al. Presence of the mu3 opiate receptor in endothelial cells: Coupling to nitric oxide production and vasodilation. J Biol Chem 1995; 270(51):30290-30293.
61. Makman MH, Dobrenis K, Downie S et al. Presence of opiate alkaloid-selective μ_3 receptors in cultured astrocytes and in brain and retina. Adv Exp Med Biol 1996; 402:23-28.
62. Rialas C, Bilfinger TV, Salzet M et al. Endomorphin 1 and 2 do not interact with the μ_3 opiate receptor subtype. Acta Pharmacol Sin 1998; 19:403-407.
63. Rialas C, Weeks BS, Cadet P et al. Nociceptin, endomorphin-1 and -2 do not interact with invertebrate immune and neural μ_3 opiate receptor. Acta Pharmacol Sinica 2000; 21(6):516-520.
64. Pasternak GW. The Opiate Receptors. New Jersey: Humana Press, 1988.
65. Makman MH, Bilfinger TV, Stefano GB. Human granulocytes contain an opiate receptor mediating inhibition of cytokine-induced activation and chemotaxis. J Immunol 1995; 154:1323-1330.
66. Fimiani C, Mattocks DW, Cavani F et al. Morphine and anandamide stimulate intracellular calcium transients in human arterial endothelial endothelial cells: coupling to nitric oxide release. Cellular Signaling 1999; 11(3):189-193.
67. Laurent V, Salzet M. Identification and properties of an angiotensin-converting enzyme in the leech *Theromyzon tessulatum*. Peptides 1996; 17(5):737-745.
68. Laurent V, Salzet M. Isolation of a neuropeptide-degrading endopeptidase from the leech *Theromyzon tessulatum*. Eur J Biochem 1995; 233:186-191.
69. Laurent V, Salzet M. Metabolism of angiotensins in head membranes of the leech *Theromyzon tessulatum* by peptidases. FEBS Lett 1996; 384:123-127.
70. Laurent V, Stefano GB, Salzet M. The leech angiotensin-converting enzyme. Ann N Y Acad Sci 1997; 839:500-502.
71. Smith EM, Hughes TK, Hashemi F et al. Immunosuppressive effects of ACTH and MSH and their possible significance in human immunodeficiency virus infection. Proc Natl Acad Sci USA 1992; 89:782-786.
72. Stefano GB, Scharrer B. A possible immunoregulatory function for [Met]-enkephalin-Arg6-Phe7 involving human and invertebrate granulocytes. J Neuroimmunol 1991; 31:97.
73. Stefano GB, Smith EM. Adrenocorticotropin, a central trigger in immune responsiveness: Tonal inhibition of immune activation. Med Hypotheses 1996; 46:471-478.
74. Salzet M, Vieau D, Stefano GB. Serpins: An evolutionarily conserved survival strategy. Immunol Today 1999; 20(12):541-544.
75. Soeparwata R, Hartman A, Frerichmann U et al. Aprotinin diminishes inflammatory processes. Int J Cardiol 1996; 53:S55-S64.

76. Chopin V, Stefano GB, Salzet M. Tessulin: a new trypsin-cathepsine G inhibitor from the leech *Theromyzon tessulatum*. Eur J Biochem 1998; 254(3):565-570.
77. Fricchione GL, Bilfinger TV, Jandorf L et al. Surgical anticipatory stress manifests itself in immunocyte desensitization: evidence for autoimmunoregulatory involvement. Int J Cardiol 1996; 53:S65-S74.
78. Shipp MA, Stefano GB, Switzer SN et al. CD10 (CALLA)/neutral endopeptidase 24.11 modulates inflammatory peptide-induced changes in neutrophil morphology, migration, and adhesion proteins and is itself regulated by neutrophil activation. Blood 1991; 78:1834-1841.
79. Stefano GB, Leung MK, Bilfinger TV et al. Effect of prolonged exposure to morphine on responsiveness of human and invertebrate immunocytes to stimulatory molecules. J Neuroimmunol 1995; 63:175-181.
80. Liu Y, Bilfinger TV, Stefano GB. A rapid and sensitive quantitation method of endogenous morphine in human plasma. Life Sci 1996; 60(3):237-243.
81. Bilfinger TV, Hartman A, Liu Y et al. Cryopreserved veins used for myocardial revascularization: a 5 year experience and a possible mechanism for their increased failure. Ann Thorac Surg 1997; 63:1063-1069.
82. Nikolarakis KE, Pfeiffer A, Stalla G et al. Facilitation of ACTH secretion by morphine is mediated by activation of CRF releasing neurons and sympathetic neuronal pathways. Brain Res 1989; 498:385-388.
83. Prevot V, Rialas C, Croix D et al. Morphine and anandamide coupling to nitric oxide stimulated GnRH and CRF release from rat median eminence: neurovascular regulation. Brain Res 1998; 790(1-2):236-244.
84. Malfroy B, Swerts JP, Guyon A et al. High affinity enkephalin degrading peptidase in brain is increased after morphine. Nature 1978; 276:523-526.
85. Genedani S, Bernardi M, Ottaviani E et al. Differential modulation of invertebrate hemocyte motility by CRF, ACTH, and its fragments. Peptides 1993; 15:203-206.
86. Brown SL, Tokuda S, Saland LC et al. Opioid peptides effects on leukocyte migration. In: Plotnikoff NP, Faith RE, Murgo AJR, eds. Enkephalins and Endorphins: Stress and the Immune System. New York: Plenum Press, 1986:367-386.
87. Brown SL, Van Epps DE. Opioid peptides modulate production of interferon gamma by human mononuclear cells. Cell Immunol 1986; 103:19-26.
88. Fischer EG, Falke NE. The influence of endogenous opioid peptides on venous granulocytes. In: Plotnikoff NP, Faith RE, Murgo AJR et al, eds. Enkephalins and Endorphins Stress and the Immune System. New York: Plenum Press, 1986:263-270.
89. Falke NE, Fischer EG. Opiate receptor mediated internalization of 125I-beta-endorphin in human polymorphonuclear leukocytes. Cell Biol Int Rep 1986; 10:429-437.
90. Stefano GB. Role of opioid neuropeptides in immunoregulation. Prog Neurobiol 1989; 33:149-159.
91. Paemen LR, Porchet-Hennere E, Masson M et al. Glial localization of interleukin-1a in invertebrate ganglia. Cell Mol Neurobiol 1992; 12:463-472.
92. Hughes TK, Smith EM, Cadet P et al. Interaction of immunoactive monokines (IL-1 and TNF) in the bivalve mollusc *Mytilus edulis*. Proc Natl Acad Sci USA 1990; 87:4426-4429.
93. Beck G, Habicht GS. Isolation and characterization of a primitive interleukin-1-like protein from an invertebrate, *Asterias forbesi*. Proc Natl Acad Sci USA 1986; 83(19):7429-7433.
94. Szucs A, Stefano GB, Hughes TK et al. Modulation of voltage-activated ion currents on identified neurons of *Helix pomatia* L. by interleukin-1. Cell Mol Neurobiol 1992; 12(5):429-438.
95. Stefano GB, Smith EM, Hughes TK. Opioid induction of immunoreactive interleukin-1 in *Mytilus edulis* and human immunocytes: an interleukin-1-like substance in invertebrate neural tissue. J Neuroimmunol 1991; 32(1):29-34.
96. Stefano GB, Cadet P, Dokun A et al. A neuroimmunoregulatory-like mechanism responding to electrical shock in the marine bivalve *Mytilus edulis*. Brain Behav Immun 1990; 4(4):323-329.
97. Gintzler AR, Levy A, Spector S. Antibodies as a means of isolating and characterizing biologically active substances: Presence of a nonpeptide morphine-like compound in the central nervous system. Proc Natl Acad Sci USA 1976; 73:2132-2136.
98. Goldstein A, Barrett RW, James IF et al. Morphine and other opiates from beef brain and adrenal. Proc Natl Acad Sci USA 1985; 82:5203-5207.
99. Cardinale GJ, Donnerer J, Finck AD et al. Morphine and codeine are endogenous components of human cerebrospinal fluid. Life Sci 1987; 40:301-306.

100. Donnerer J, Cardinale G, Coffey J et al. Chemical characterization and regulation of endogenous morphine and codeine in the rat. J Pharmacol Exp Ther 1987; 242:583-587.
101. Donnerer J, Oka K, Brossi A et al. Presence and formation of codeine and morphine in the rat. Proc Natl Acad Sci USA 1986; 83:4566-4567.
102. Oka K, Kantrowitz JD, Spector S. Isolation of morphine from toad skin. Proc Natl Acad Sci USA 1985; 82:1852-1854.
103. Stefano GB, Scharrer B. Endogenous morphine and related opiates, a new class of chemical messengers. Adv Neuroimmunol 1994; 4:57-68.
104. Sonetti D, Mola L, Casares F et al. Endogenous morphine levels increase in molluscan neural and immune tissues after physical trauma. Brain Res 1999; 835(2):137-147.
105. Goumon Y, Casares F, Zhu W et al. The presence of morphine in ganglioic tissues of *Modiolus deminissus*: A highly sensitive method of quantitation for morphine and its derivatives. Mol Brain Res 2000; 86:184-188.
106. Stefano GB, Goumon Y, Casares F et al. Endogenous morphine. Trends Neurosci 2000; 9:436-442.
107. Goumon Y, Casares F, Pryor S et al. *Ascaris suum*, an internal parasite, produces morphine. J Immunol 2000; 165(1):339-343.
108. Bianchi E, Alessandrini C, Guarna M et al. Endogenous codeine and morphine are stored in specific brain neurons. Brain Res 1993; 627:210-215.
109. Bianchi E, Guarna M, Tagliamonte A. Immunocytochemical localization of endogenous codeine and morphine. Adv Neuroimmunol 1994; 4:83-92.
110. Leung MK, Dissous C, Capron A et al. *Schistosoma mansoni*: The presence and potential use of opiate-like substances. Exp Parasit 1995; 81:208-215.
111. Laurent V, Salzet B, Verger-Bocquet M et al. Morphine-like substance in leech ganglia. Evidence and immune modulation. Eur J Biochem 2000; 267(8):2354-2361.
112. Ferreira SH, Duarte ID, Lorenzetti BB. The molecular mechanism of action of peripheral morphine analgesia: stimulation of the cGMP system via nitric oxide release. Eur J Pharmacol 1991; 201:121.
113. Przewlocki R, Machelska H, Przewlocka B. Inhibition of nitric oxide synthetase enhances morphine antinociception in the rat spinal cord. Life Sci 1993; 53:PL1-5.
114. Majeed NH, Przewlocka B, Machelska H et al. Inhibition of nitric oxide synthase attenuates the development of morphine tolerance and dependence in mice. Neuropharmacol 1994; 33:189.
115. Fecho K, Maslonek KA, Coussons-Read ME et al. Macrophage-derived nitric oxide is involved in the depressed concanavalin A responsiveness of splenic lymphocytes from rats administered morphine in vivo. J Immunol 1994; 152:5845-5851.
116. Gyires K. The role of endogenous nitric oxide in the gastroprotective action of morphine. Eur J Pharmacol 1994; 255:33.
117. Sawada M, Ichinose M, Stefano GB. Nitric oxide inhibits the dopamine-induced K+ current via guanylate cyclase in Aplysia neurons. J Neurosci Res 1997; 50(3):450-456.
118. Stefano GB, Salzet B, Rialas CM et al. Morphine and anandamide stimulated nitric oxide production inhibits presynaptic dopamine release. Brain Res 1997; 763:63-68.
119. Bilfinger TV, Fimiani C, Stefano GB. Morphine's immunoregulatory actions are not shared by fentanyl. Int J Cardiol 1998; 64(S1):61-66.
120. Stefano GB. Autoimmunovascular regulation: morphine and anandamide stimulated nitric oxide release. J Neuroimmunol 1998; 83:70-76.
121. Stefano GB, Salzet M, Bilfinger TV. Long-term exposure of human blood vessels to HIV gp120, morphine and anandamide increases endothelial adhesion of monocytes: Uncoupling of Nitric Oxide. J Cardiovasc Pharmacol 1998; 31:862-868.
122. Stefano GB, Salzet M, Magazine HI et al. Antagonist of LPS and IFN-α induction of iNOS in human saphenous vein endothelium by morphine and anandamide by nitric oxide inhibition of adenylate cyclase. J Cardiovasc Pharmacol 1998; 31:813-820.
123. Rojavin M, Szabo I, Bussiere JL et al. Morphine treatment in vitro or in vivo decreases phagocytic functions of murine macrophages. Life Sci 1993; 53(12):997-1006.
124. Szabo C. Alterations in nitric oxide production in various forms of circulatory shock. New Horizons 1995; 3:2-32.
125. Szabo C. Physiological and pathophysiological roles of nitric oxide in the central nervous system. Brain Res Bull 1996; 41(3):131-141.
126. Tubaro E, Borelli G, Croce C et al. Effect of morphine on resistance to infection. J Infect Dis 1983; 148:656-666.

127. Weber RJ, Band LC, DeCosta B et al. Neural control of immune function: Opioids, opioid receptors and immunosuppression. NIDA Res Monogr 1991; 105:96-102.
128. Perez-Castrillon JP, Perez-Arellano JP, Garcia-Palomo JD et al. Opioids depress in vitro human monocyte chemotaxis. Immunopharmacol 1992; 23:57-61.
129. Stefano GB, Bilfinger TV. Human neutrophil and macrophage chemokinesis induced by cardiopulmonary bypass: Loss of DAME and IL-1 chemotaxis. J Neuroimmunol 1993; 47:189-198.
130. Stefano GB, Hiripi L, Rozsa KS et al. Behavioral effects of morphine on the land snail *Helix pomatia*: Demonstration of tolerance. In: Salanki J, ed. Neurobiology of Invertebrates. New York: Pergamon Press, 1980:285-295.
131. Kavaliers M, Hirst M, Teskey GC. A functional role for an opiate system in snail thermal behavior. Science 1983; 330:99-103.
132. Stefano GB, Smith DM, Smith EM et al. MSH can deactivate both TNF stimulated and spontaneously active immunocytes. In: Kits KS, Boer HH, Joosse J, eds. Molluscan Neurobioogy. Amsterdam: North Holland Publishing Company, 1991:206-209.
133. Liu Y, Casares F, Stefano GB. D2 opioid receptor mediates immunocyte activation. Chinese J Neuroimmunol Neurol 1996; 3(2):69-72.
134. Stefano GB, Casares F, Liu Y. Naltrindole sensitive d2 opioid receptor mediates invertebrate immunocyte activation. Acta Hungaria 1995; 321-327.
135. Stefano GB. The μ_3 opiate receptor subtype. Pain Forum 1999; 8:206-209.
136. Goumon Y, Stefano GB. Identification of Morphine in the Rat Adrenal Gland. Mol Brain Res 2000; 77:267-269.
137. Nieto-Fernandez FE, Mattocks DW, Cavani F et al. Morphine coupling to invertebrate immunocyte nitric oxide release is dependent on intracellular calcium transients. Comp Biochem Physiol 1999; 123(3):295-299

CHAPTER 12

Anti-Inflammatory Effects of Opioids

Judith S. Walker

Overview

Rheumatoid arthritis (RA) is a chronic systemic inflammatory disorder with its primary manifestations in the joints. The etiology of RA remains obscure, no cure is yet available and sustained disease remission is rarely achieved. Opioid drugs are not currently used in the treatment of RA, partly because of their range of side-effects and because their anti-inflammatory (as opposed to analgesic) actions have been largely unrecognized. Analgesic compounds with some central κ-agonist activity, such as pentazocine and butorphanol have been available clinically for a number of years for treatment of pain but they have not been utilized extensively due to their dysphoric side effects. The synthesis of peripherally selective κ-opioid agonists has allowed the analgesic and anti-inflammatory effects of opioids in arthritis to be studied, while mitigating the problems of tolerance and central side effects. They are powerfully anti-inflammatory in a dose-dependent, time-dependent, stereoselective and antagonist reversible manner.[1] This chapter examines the anti-inflammatory effects of κ-opioids, both centrally active and peripherally selective κ-opioid agonists, with particular relevance to RA, and reports data on the mechanisms responsible for the anti-arthritic effects of κ-opioids in adjuvant arthritis.

Opioids—Receptor Pharmacology

Opioids exert their diverse physiological effects through three distinct membrane-bound receptor subtypes mu (μ), delta (δ) and kappa (κ) in the CNS[2] and periphery.[3] The opioid receptors mediate the antinociceptive (analgesic) and other pharmacological (such as respiratory, cardiovascular) actions of opioid drugs. They also regulate responses to pain, stress and emotions when activated by endogenous opioid peptides. The three members of the opioid receptor were cloned in the early 1990s, belong to the family of seven-transmembrane G-protein coupled receptors[4] and are highly homologous (60-90%).[5] A plethora of studies have reviewed their properties[2,6] and their distribution throughout the central and peripheral nervous systems.[7,8] The different receptors have diverse behavioral characteristics[2] for example, euphoria, physical dependence and respiratory depression are mainly associated with μ and δ receptors. In contrast, opioids acting at κ-receptors produce dysphoric rather than euphoric effects which limits their physical dependence liability.[9,10] In this regard, κ-opioid agonists possess some advantages over μ-agonists: they are devoid of such side effects as dependence liability, constipation and respiratory depression.[9,11]

Immune Mechanisms of Pain and Analgesia, edited by Halina Machelska and Christoph Stein.

Anti-Inflammatory Effects of Opioids

Classically, opioids have been used in the treatment of pain rather than inflammation, partly due to their side effects and because their anti-inflammatory actions have been largely unrecognized. A great deal is known about the analgesic effects of opioids[12] and the actions of opioids on the hyperalgesic aspects of inflammation have been comprehensively reviewed.[13] Apart from our own work there have been relatively few studies of their peripheral anti-inflammatory effects so a brief overview of these effects is presented here.

There are conflicting reports as to whether μ-opioids have anti- or pro-inflammatory properties.[14] For example, morphine inhibits carrageenan-induced paw swelling[15] and near toxic doses of morphine were able to attenuate adjuvant arthritis in rats:[16,17] By contrast, low doses of morphine were pro-inflammatory in adjuvant arthritis.[18] High doses would preclude clinical use of morphine in arthritis, so specific attention was given to κ-opioids, particularly since they have a more favorable side effect profile.

Opioids—Peripheral Actions

Historically, opioids have been thought to produce their analgesic effects via actions in the central nervous system (CNS), but it is now well appreciated that opioid receptors are synthesized in the dorsal root ganglia and transported towards both central and peripheral nerve terminals. Further, the peripheral axonal transport is upregulated during inflammation (see ref. 19). Experimental and clinical studies have shown potent analgesic effects after peripheral administration of opioids (see ref. 20). For example, the pioneering work of Robert Schmidt's group in Germany has shown the local action of opioids in the knee joint of the cat.[21] Morphine (μ-agonist) and PNU50488H (κ-agonist) reduce the action potential frequency in group III (Aδ) fibres of an articular nerve; reversal of this action by naloxone indicates an opioid receptor mediated action.[21] This local action has clinical significance as intra-articular morphine produced pain relief following knee arthroscopy[22] and in patients undergoing dental surgery after submucous injection without overt systemic effects[22] (for review see refs. 20, 23). A large body of work has also demonstrated that local administration of low doses of opioid receptor agonists elicit potent analgesic effects in inflamed but not noninflamed tissue (for review see refs. 14, 24-26). Clearly, there are functional opioid receptors on the peripheral terminals of afferent nerves which could well be exploited clinically.

κ-Opioids

Kappa-agonists belong to four chemical classes, namely: the peptides (related to the endogenous ligand dynorphin), the benzomorphans (prototype ethylketocyclazocine), the arylacetamides (prototype PNU50488H), and the benzodiazepine derivative tifluadom. Although they all have some affinity for the μ and δ-opioid receptors, the arylacteamides have been found to be the purest κ opioids, with binding affinities in the nanomolar range.[27,28] They include the centrally acting and prototype PNU50488H which has served as the structural starting point for the synthesis of a multitude of compounds such as the centrally acting compounds: PNU62066E, PD117302, GR89696A.

To minimize the problems of tolerance and central side effects various chemical approaches have been utilized to make opioids less accessible to the brain without reducing κ-opioid activity. The structure of PNU50488H has been the basis for the development of many of these compounds.[28] Asimadoline (EMD 61753; Merck KGaA), an amphiphilic compound which is orally active, is undergoing phase II clinical trials against musculoskeletal pain.[28,29] Most recently, other arylacetamide derived peripherally selective agents ADL 10-0101 and ADL-10-0116 (Adolor Corporation) have been utilized in animal studies. Others include ICI 204448, ICI 197067, GR94839, EMD 60400, fedotozine.[28-32]

κ-Opioids as Anti-Arthritic Agents

In our laboratory we induce chronic polyarthritis in the rat by administration of complete Freund's adjuvant.[17-33] We measure disease severity using three quantitative indicators – paw swelling (edema), radiological damage and histological characteristics. All three measures increase in severity as disease progresses with an apparent plateau after day 21 except for radiology which continues to progress beyond day 28.[33] As might be expected, these joints are painful; there is a 30% reduction in the mechanical force required for a paw withdrawal threshold i.e., hyperalgesia is evident.[34] We have also used immunohistochemistry to demonstrate that both mast cells and macrophages increase in number throughout the process, with the mast cells reaching a plateau from about 13 days. We went on to test the anti-inflammatory effects of κ opioids in this model. Here the κ-opioids, (e.g., PNU50488H and asimadoline) attenuated the progression of experimental adjuvant arthritis via specific opioid receptors in the periphery using rigorous criteria such as reversibility by opioid antagonists, dose-dependency and stereospecificity.[1,14,17,33,35] For example, all the indicators of disease severity were reduced by asimadoline (Fig. 1A) and the tissue populations of inflammatory cells are also reduced by as much as 80% (Fig. 1B). This effect was seen with a number of κ-opioids, including low doses of centrally acting compounds PNU50488H administered peripherally into a joint[33,36] and peripherally selective asimadoline[1] so our work has clinical potential because these drugs could be used with minimal likelihood of central side effects such as addiction or tolerance. By contrast, morphine was only anti-inflammatory at very high doses (Table 1; ED50 ~ 58 mg/kg).

The temporal details of the treatment regimens were found to be important. The opioid action is most significant in the first few days of treatment (i.e., disease onset). As shown in Fig. 1A, treatment with asimadoline (5mg/kg/day) significantly attenuated adjuvant arthritis provided it was administered during the period of disease onset i.e., in animals treated over the first three days or over the entire period (days 1-21). When the animals received drug during established disease (days 13-21), improvement was only observed in the histology assessment. These data support current opinion that aggressive drug therapy needs to be started as soon as possible after disease onset to prevent progressive joint destruction.[1,33,37]

Table 1. Summary of ED50's (dose at half maximal anti-inflammatory effect; mg/kg) for selected opioid agonists

Agonist	ED50 (mg/kg)
κ-class	
PNU50488H	19 ± 1
PD117302	14 ± 7
Asimadoline	1.3 ± 0.1
μ-class	
Morphine	60 ± 9

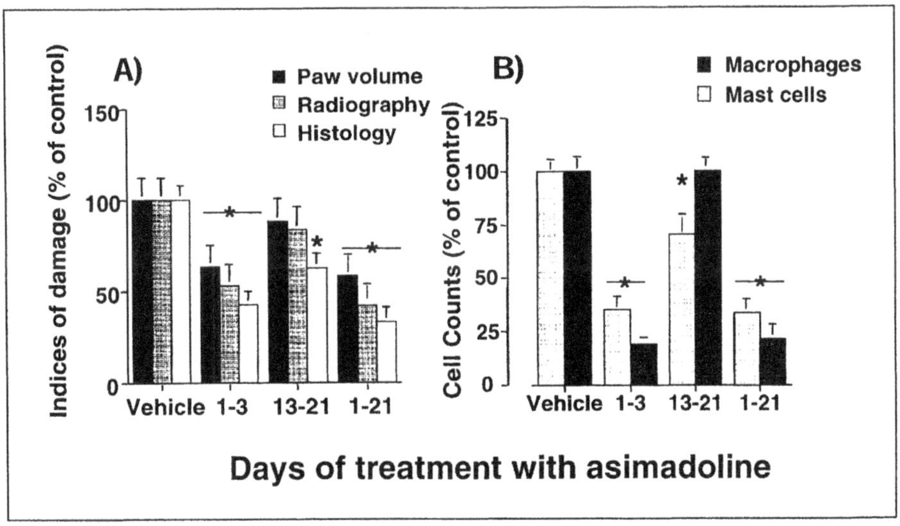

Fig. 1. Effects of asimadoline or vehicle administration on A) indices of arthritis severity: paw volume, radiography and histology or B) on the reduction in either macrophage or mast cell numbers after 5 mg/kg/day intraperitoneal injection as a function of treatment time. Data are expressed as percentage of vehicle-treated control rats (100%). * Denotes $p<0.05$ compared to vehicle treated controls. See text for details.

The anti-arthritic actions of κ-opioids were blocked by the opioid antagonist naloxone methiodide, which does not penetrate the CNS; and by specific κ-opioid antagonists, thus, the anti-inflammatory effects are most likely exerted in the periphery via κ-opioid receptors.

The higher incidence of inflammatory and painful disorders in women and recent reports which have emphasized the importance of gender in nociceptive sensitivity (for review see ref. 38) and responsiveness to analgesics[39,40] (for review see ref. 38) prompted us to investigate gender as a factor in the variability in response to opioids. In humans, females with postoperative dental pain responded better to the analgesic actions of κ-opioids (for review see refs. 38, 40). We found that both the centrally acting PNU50488H and the peripherally selective compound asimadoline had equally powerful anti-inflammatory effects in both male and female rats (reducing measures by 60-80%). By contrast, there were gender-based heterogeneities in their analgesic actions, contingent upon the method of stimulation (mechanical or thermal); males were insensitive to the analgesic effects of asimadoline using thermal but not mechanical nociceptive stimuli.[34] The gender based heterogeneities provide a significant advantage for κ opioids given the female preponderance in inflammatory diseases.

Mechanisms Responsible for Anti-Inflammatory Effects of Opioids

The finding that κ opioids reduce inflammation in adjuvant arthritis raises interesting questions about the mechanisms and inflammatory mediators involved. The actions of κ-opioids are mediated via κ-receptors, which are found on immune system cells and on neural cells (Fig. 2; for review see ref. 41 and chapters 5, 8). The close spatial and functional association between nerves and immune cells suggests modulation of immune

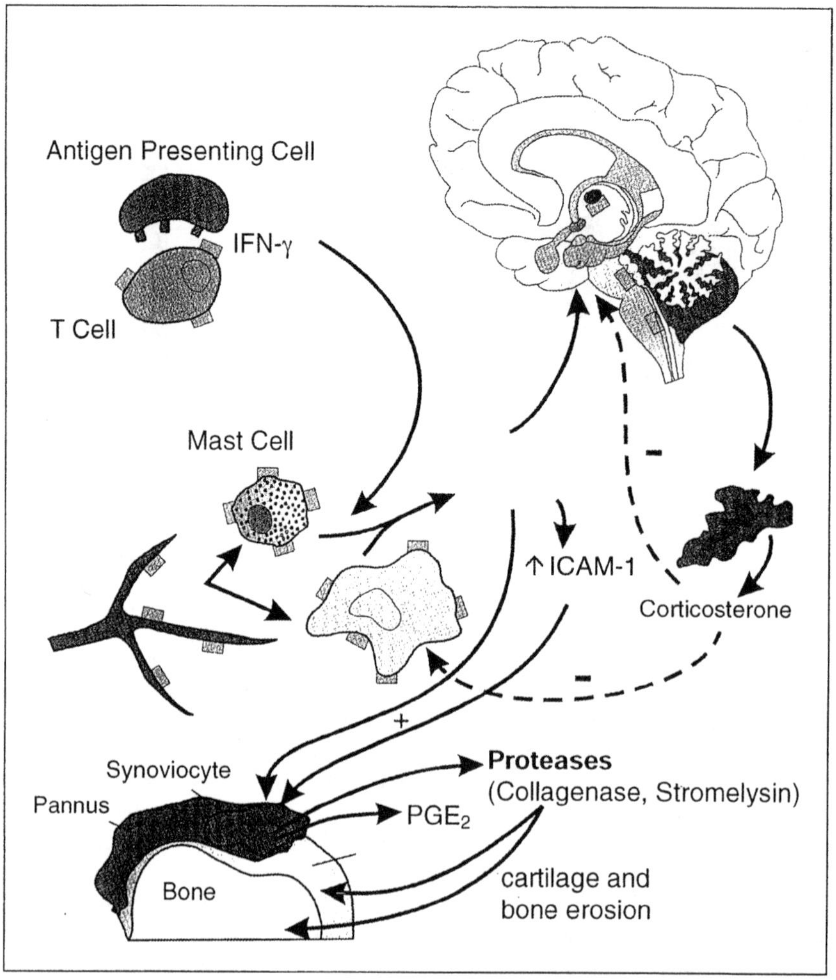

Fig. 2. Schematic illustration of neuro-immune interactions in the control of inflammation. Opioid receptors are identified by the purple blocks and suggest possible sites of action. INF-γ, interferon-γ; ICAM-1, intercellular cell adhesion molecule-1; PGE_2, prostaglandin E_2.

or neural pro-inflammatory pathways could contribute to the disease suppressing activity of these drugs.[14]

Hypothalamic-Pituitary-Adrenal (HPA) Axis

A major point of neuro-immune convergence is in what is known as the HPA axis. It has been proposed that dysfunction of this axis may contribute to the susceptibility to, and even the persistence of, human RA.[42,43] In general, this system is controlled by negative feedback of glucocorticoids but its sensitivity can be influenced by opioids as well as by cytokines[44-46] (Fig. 3). Is it possible therefore that the anti-arthritic action of the κ-opioids is mediated via the HPA axis? Using the prototype PNU50488H in adrenalectomized rats, arthritis developed sooner and was more severe. However, PNU-50488H substantially

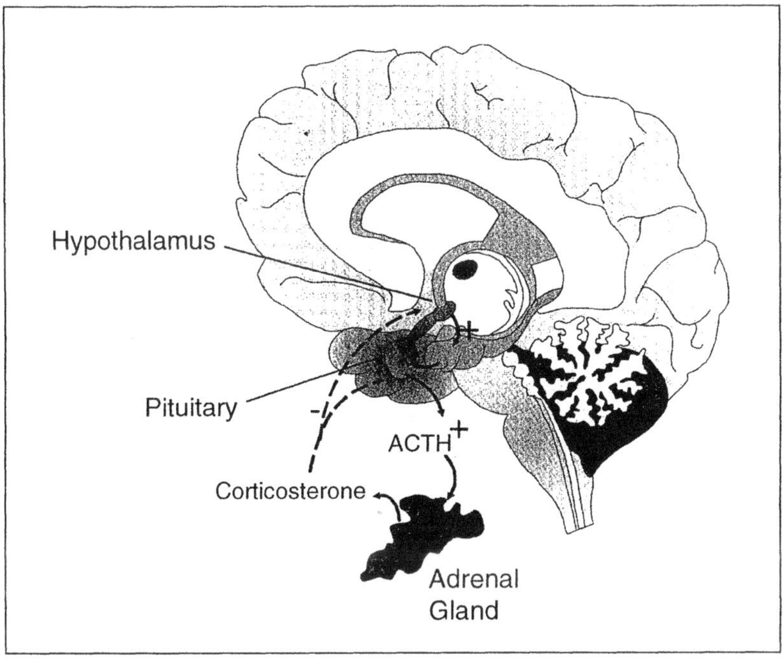

Fig. 3. The role of the hypothalamic-pituitary (HPA) axis in inflammation. An inflammatory stimulus initiates the release of cytokines from immune cells which in turn stimulate the hypothalamus to secrete corticoreleasing factor (CRF) which is then transported to the pituitary gland. In turn, CRF stimulates the release of adrenocorticotrophic hormone (ACTH) from the pituitary. Circulating ACTH then stimulates the release of glucocorticoids from the adrenal gland. Glucocorticoids produce a negative feedback on the HPA axis by inhibiting basal and stimulated secretion of ACTH and CRF as well as cytokines from immune cells.

reduced the pooled severity index (combined quantitative oedema, histological and radiological assessments) at day 18 in both control and adrenalectomized rats to an equal extent. Thus, the HPA-axis is only partially involved in the anti-inflammatory actions of opioids.[34] We therefore continued our investigations into other neural and immune mechanisms.

Role of the Nervous System

Neuropeptides such as substance P (SP) and calcitonin gene-related peptide (CGRP) are strongly implicated in the pathogenesis and/or spread of inflammatory arthritis.[47] Synovial joints are extensively innervated by afferent fibers, which contain both SP and CGRP (for review see ref. 48).

By using two differently acting drugs (the κ-agonist asimadoline, and the NK_1 antagonist GR205171) the roles that SP might play in the pathogenesis and maintenance of experimental arthritis were investigated. Dependent upon the timing of their administration, both agents significantly attenuated adjuvant arthritis, putatively via a mechanism that involves SP: we hypothesize that asimadoline acts on peripheral terminals to modulate SP release, while GR205171 antagonizes the action of SP, either peripherally or centrally. Time-dependent multiphasic effects with asimadoline on SP tissue levels (i.e., protein) in the joints from rats with adjuvant arthritis were found; treatment with asimadoline decreased SP levels in the joint during early arthritis (day 3) and increased levels were found during established disease (day 21).[34] Drug-induced changes in SP content could result from change in the release or synthesis from neural cells. We therefore further investigated neuropeptide production in cell bodies innervating the joints as a function of time. Asimadoline reduced expression of both SP and CGRP mRNA in the DRG on either day 13 or day 21 post-adjuvant[49] (Fig. 4) This reduction in peptide mRNA is likely to be a consequence of a negative feedback mechanism resulting from an increase in SP in the nerve terminals (and therefore enhanced axonal transport) rather than a direct action on peptide synthesis by the κ-agonist.[49] Our studies confirm literature findings on the inhibitory actions of opioids on SP release from the peripheral endings of primary afferent fibers (for review see ref. 48).

We also sought evidence for gender influences on the joint content of SP and found that levels were higher in the untreated arthritic females. However, there were no gender differences in disease severity or nociception sensitivity in untreated arthritic animals. Paradoxically, asimadoline and GR205171 elevated SP concentrations in the joints, perhaps as a consequence of an action of κ-opioids to suppress SP release from peripheral nerves, but the gender differences remained. Further experiments are required to determine exact mechanisms responsible for the gender distinction in analgesic response to k-opioids which may involve differential activation of primary afferents.

Role of the Immune System

Opioids have been found to regulate lymphocyte proliferation, antibody production and natural killer cell activity as well as inhibiting the function of neutrophils, monocytes and macrophages (for review see ref. 50 and chapter 9). Potent peripheral analgesia can also be induced by interaction of peripheral opioid receptors with endogenous opioid peptides released from immune cells under inflammatory conditions.[51] Thus the hypothesis that opioids might be exerting their anti-inflammatory actions via opioid receptors that exist on immune cells as shown by Sharps' group (see ref. 41 and chapter 8), together with our own observations that opioids decrease the numbers of both mast cells and macrophages infiltrating the joint (Fig. 1B) is an attractive one.

Extravasation of immune cells is a multistep process involving the sequential activation of adhesion molecules on immune cells and the vascular endothelium. Interruption of the leucocyte-endothelial cell cascade by antibodies against adhesion molecules blocks immune cell extravasation.[52] Furthermore, pretreatment of rats with a selectin blocker, fucoidin, abolishes peripheral opioid analgesia (see ref. 53 and chapter 7). The question arises whether adhesion molecule blockade also abolishes inflammation in arthritis.

Effect on Adhesion Molecules

Intercellular adhesion molecule-1 (ICAM-1) is critical for recruitment of leucocytes into the affected joints and a 60% increase in the number of ICAM-1 positive cells in the endothelium has been reported in patients with rheumatoid arthritis.[54] Therefore modifying the expression of ICAM-1 has the potential to modulate inflammation in the joint. The effect of the κ-opioid PD 117302 on ICAM-1 expression was thus compared with the effects of a prototypic nonsteroidal anti-inflammatory drug, naproxen, in the same model. ICAM-1 expression was significantly upregulated in the joints of affected limbs of animals with both unilateral hind paw inflammation and polyarthritis.[55] In animals treated with PD 117302 and naproxen there was a significant attenuation of arthritis, however, only treatment with PD 117302 was able to significantly inhibit the upregulation of ICAM-1 expression in arthritic joints[55] (Fig. 5A,B).

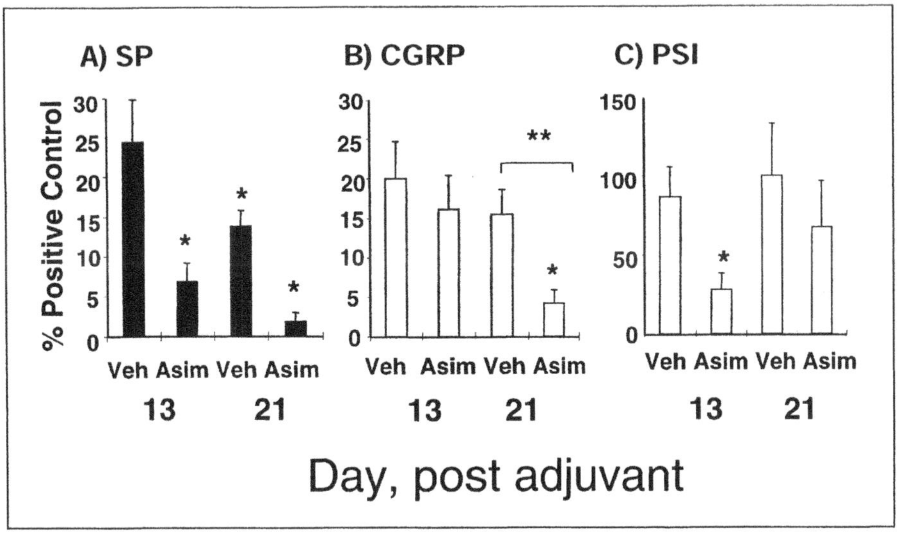

Fig. 4. Effect of asimadoline (Asim) compared to saline on the expression of neuropeptides; A) SP; B) CGRP in DRG neurons from L4-L6 or C) time course of the development of arthritis as judged by a pooled severity index (PSI) summing paw swelling, radiography and histology to give a comprehensive measure of disease severity on either day 13 or day 21 post-adjuvant. *$P<0.05$ compared with saline treated rats, (n= 6-8). SP, substance P; CGRP, calcitonin gene-related peptide; DRG, dorsal root ganglia; L4-L6, lumbar DRG 4-6. For details see text.

Inhibition of adhesion molecule expression by κ-opioids could occur via either a direct effect on endothelial cells; or an indirect effect via inhibition of release of its regulators such as tumor necrosic factor (TNF) from immune cells. In support of the latter, the upregulation of ICAM-1 expression in arthritic joints, is thought to be stimulated by pro-inflammatory cytokines including interleukin (IL)-1, TNF and interferon (IFN)-γ which are released from inflammatory cells such as macrophages.[54,56,57] Kappa-opioids also directly inhibit the release of both IL-1 and TNF from macrophages.[58] Decreased cytokine release may explain the abrogation of the upregulation of ICAM-1 expression, reduced leucocyte recruitment and less synovial inflammation observed in the adjuvant arthritic rats. Others have found that anti-TNF therapy in adjuvant arthritis also decreases inflammation and leucocyte recruitment to joints.[59] We further investigated this using cultured macrophages in vitro and examined the effects of range of κ-opioids on TNF release from these cells.

Effect of κ-Agonists on TNF Release from Macrophages

The production of TNF was significantly inhibited in the presence of asimadoline, PD117302, PNU50488H in a dose-dependent manner over the dose range of 10^{-11} to 10^{-3} M with 100% suppression at a concentration of 10^{-3} M (see Fig. 6 for results with asimadoline, ED50: 9×10^{-7}). Although the specific κ-antagonist nor-binaltorphimine (NORBNI), at a dose of 10^{-4} M, could not inhibit the highest dose of agonist, it could block their effects at other doses (Fig. 6). By contrast, morphine (a μ-opioid) and SNC80 (a δ-opioid) were without effect in this system. Thus, κ-opioids may be acting directly to inhibit cytokine release from immune cells. Taken together with our earlier findings of decreased cell recruitment and reduced expression of adhesion molecules, these data help

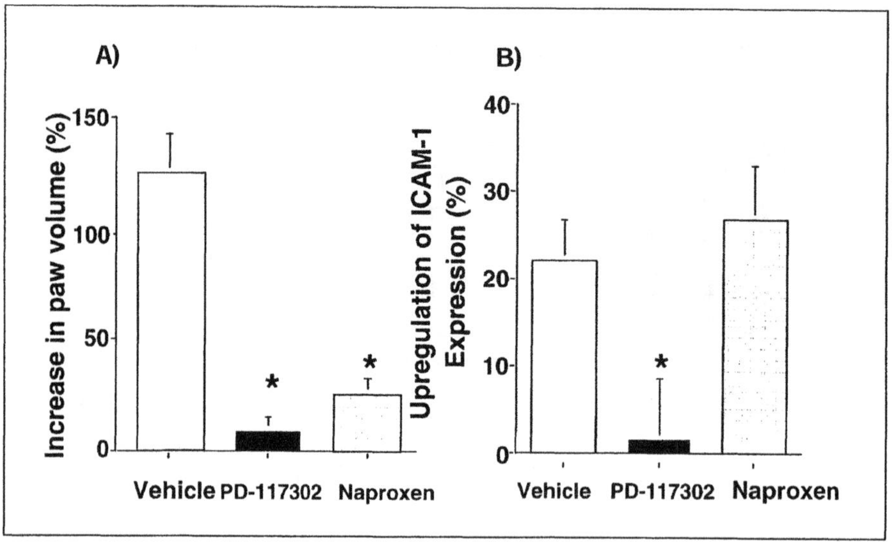

Fig. 5. Effect of treatment with PD 117302 or naproxen on (A) polyarthritis as measured by hind paw edema expressed as the percent increase in paw volume of both hind paws (value for normal rats was 4.5 ± 7.3%), and (B) the % increase in ICAM-1 expression above basal values in the ankle joints (basal levels for normal rats was 0.6 ± 0.002 %). *P<0.05 represents significant attenuation compared with vehicle treated controls (n=9-10). ICAM-1, intercellular adhesion molecule-1.

explain the powerful anti-inflammatory effects observed clinically in the adjuvant arthritis model. Studies are now being extended to human synovium.

Effect on Cytokine mRNA Expression

Our results to date show that asimadoline reduces synovial macophage and mast cell numbers in adjuvant arthritis. In addition, it suppresses TNF production by lipopolysaccharide-stimulated macrophages in vitro.[36] Taken together with the work of Belkowski et al,[58] who demonstrated that the prototype κ-opioid PNU50488H inhibits gene transcription as well as production of TNF and IL-1, we extended our studies to explore the effects of asimadoline on in vivo cytokine expression. Asimadoline significantly decreased arthritis severity by day 13, with a concomitant decrease in synovial membrane expression of cytokines, TNF, IL-17 and no change in T cell numbers in the joints of arthritic rats (Fig. 7). Interestingly, there were time-dependent changes in transforming growth factor (TGF)-β which is present in human synovium and has both pro- and anti-inflammatory actions.[59] Furthermore, asimadoline decreased intra-articular TGF-β whilst increasing extra-articular (inguinal lymph node) expression. An altered balance, therefore, in the pro- and anti-inflammatory effects of TGF-β by asimadoline might explain its striking anti-arthritic actions.

Summary and Conclusions

In summary, therapy with opioids is an exciting new development for arthritis especially since there is the potential for fewer side effects from molecules which act outside the CNS. We found κ-opioid drugs to be powerfully anti-inflammatory, reducing disease severity by as much as 80%; attenuating arthritis in a dose-dependent, stereoselective,

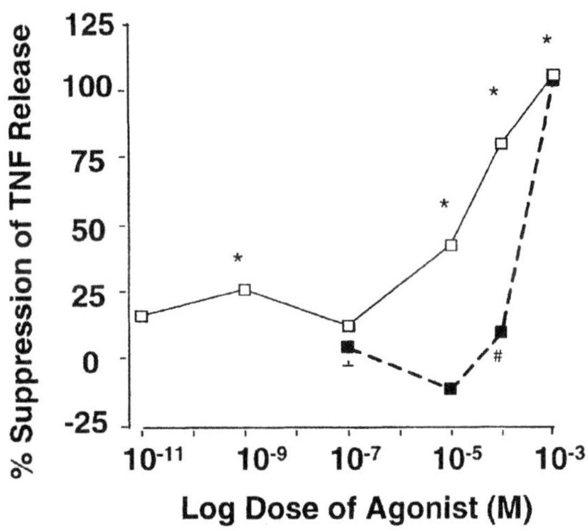

Fig. 6. The effect of the κ-opioid agonist, asimadoline, alone (□) or in the presence of the antagonist nor-binaltorphimine (■; 10^{-4} M) on the % suppression of TNF (tumor necrosis factor-α from cultured peritoneal macrophages after stimulation with 0.1 μg/ml lipopolysaccharide. * Denotes $p<0.05$ compared to untreated control. # Denotes $p<0.05$ compared to agonist alone.

antagonist-reversible manner. By contrast opioids acting at other receptors were only therapeutic at near toxic doses. The HPA-axis was found to be only partially involved, thus we investigated other neural and immune mechanisms. Results showed that the κ-opioid anti-inflammatory actions were exerted via (i) reduced adhesion molecule expression; (ii) inhibition of cell trafficking; (iii) reduced TNF release and expression and (iv) alterations in mRNA expression and protein levels of SP and CGRP in joint tissue (Fig. 2). The ability of κ-opioids to act at multiple sites in the inflammatory cascade, as suggested by the presence of opioid receptors at various locations throughout the cascade, may explain their powerful actions (Fig. 2). It is also relevant that during inflammatory states that enhanced peripherally directed axonal transport leads to receptor upregulation on peripheral nerve terminals in the joint. Neuropeptides (SP and CGRP) were found to be involved in the later phases of adjuvant arthritis suggesting that they are involved in the maintenance or persistence of the disease. The involvement of SP and the efficacy of neurokinin-1 (SP receptors) antagonists predicts that combined opioid- neurokinin-1 therapy has promise. Kappa-opioids are, however, powerfully therapeutic during disease onset. Thus, they most likely exert their anti-inflammatory effects via changes in cellular activation and cytokine expression. The mechanisms involved are summarized in Fig. 2. The increased potency of κ-opioids in females is likely to be a significant advantage for treatment of inflammatory disease with these agents. Thus our work supports the findings of Stein's group, that opioids do indeed have powerful actions in the periphery via specific receptors at that site.[19] Peripherally acting opioids may prove to be a potent new treatment for rheumatoid arthritis sufferers in the future.

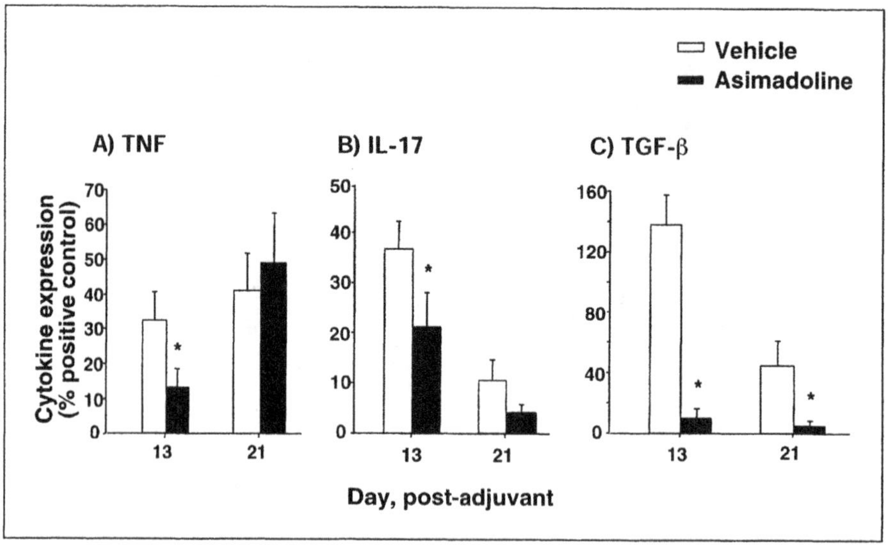

Fig. 7. Cytokine mRNA expression (as a percentage of positive control) for A) TNF, B) IL-17, and C) TGF-β in the synovium of rats with adjuvant arthritis treated with either 5mg/kg/day asimadoline (filled panels) or vehicle (open panels) i.p. * represents a significant difference from vehicle treated rats, $P<0.05$ ($n = 6\text{-}8$). TNF, tumor necrosis factor; IL-17, interleukin-14; TGF-β, transforming growth factor-β.

Acknowledgments

I am very grateful to the dedicated students, postdoctoral fellows and entire research staff of my laboratory over the past decade. In particular, I would like to thank the following persons for their contributions to this research: Dr Bruce Kirkham, A/Prof Carmody, Professor Richard Day, Drs Binder, Bush, Wilson and Scott.

References

1. Binder W, Walker JS. Effect of the peripherally selective κ-opioid agonist, asimadoline, on adjuvant arthritis. Br J Pharmacol 1998; 124:647-654.
2. Law PY, Loh HH. Regulation of opioid receptor activities. J Pharmacol Exp Ther 1999; 289:607-624.
3. Minamai M, Statoh M. Molecular biology of the opioid receptors: structures, functions and distributions. Neurosci Res 1995; 23:121-145.
4. Reisine T, Bell GL. Molecular biology of opioid receptors. Trends Neurosci 1993; 16:506-510.
5. Knapp R J, Malatynska E, Collins N et al. Molecular biology and pharmacology of cloned opioid receptors. FASEB J 1995; 9:516-525.
6. Akil H, Owens C, Gutstein H et al.. Endogenous opioids: overview and current issues. Drug Alcohol Depend 1998; 51:127-140.
7. Coggleshall RE, Zhou S, Carlton SM. Opioid receptors on peripheral sensory axons. Brain Res. 1997; 764:126-132.
8. Mansour A, Fox CA, Burke S et al. Immunohistochemical localization of the cloned m opioid receptor in the rat CNS. J Chem Neuroanat 1995; 8:283-305.
9. Horwell DC. Kappa opioid analgesics. Drugs Future 1988; 13:1061-1070.
10. Wolleman M, Benyhe S, Simon J. The kappa opioid receptor: evidence for the different subtypes. Life Sci 1996; 52:599-611.
11. Porreca F, Mosberg HT, Hurst R et al. Roles of mu, delta and kappa opioid receptors in spinal and supraspinal mediation of gastrointestinal transit effects and hotplate analgesia in the mouse. J Pharmacol Exp Ther 1984; 230:341-348.

12. Reisine T, Pasternak G. Opioid analgesics and antagonists. In: Hardman JG, Limbird LE, Molinoff PB et al, eds. The Pharmacological Basis of Therapeutics. 9th ed. New York: McGraw-Hill, 1996:521-556.
13. Barber A, Gottschlich R. Opioid agonists and antagonists: an evaluation of their peripheral actions in inflammation. Med Res Rev 1992; 12:525-562.
14. Walker JS, Wilson J, Binder W et al. The Anti-inflammatory effects of opioids: their possible relevance to the pathophysiology and treatment of rheumatoid arthritis. Res Alerts Rheum Arthritis 1997; 1:291-299.
15. Gyires K, Budavari I, Furst S et al. Morphine inhibits the carrageenan-induced edema and the chemiluminescence of leucocytes stimulated by zymosan. J Pharm Pharmacol 1985; 37:100-104.
16. Levine JD, Moskowitz MA, Basbaum AI. The contribution of neurogenic inflammation in experimental arthritis. J Immunol 1985; 155:843s-847s.
17. Walker JS, Chandler AK, Wilson JL et al. Effect of μ-opioids morphine and buprenorphine on the development of adjuvant arthritis in rats. Inflamm Res 1996; 45:557-563.
18. Earl JR, Claxson AW, Blake DR et al. Proinflammatory effects of morphine in the rat adjuvant arthritis model. Int J Tiss React 1994; 16:163-170.
19. Machelska H, Binder W, Stein C. Opioid Receptors in the Periphery. In: Kalso E, McQuay HJ, Wiesenfield-Hallin Z, eds. Opioid Sensitivity of Chronic Noncancer Pain. Progress in Pain Research and Management. Vol 14. Seattle: IASP Press, 1999.
20. Likar R, Sittl R, Gragger K et al. Peripheral morphine analgesia in dental surgery. Pain 1998; 76:145-150.
21. Russell NSW, Schaible HG, Schimdt RF. Opiates inhibit the discharges of the afferent units from inflamed knee joint of the cat. Neurosci Lett 1987; 76:107-112.
22. Stein C, Comisel K, Haimerl E et al. Analgesic effect of intraarticular morphine after arthroscopic knee surgery. N Engl J Med 1991; 325:1123-1126.
23. Kalso E, Tramer MR, Carroll D et al. Pain relief from intra-articular morphine after knee surgery: a qualitative systematic review. Pain 1997; 71:127-134.
24. Stein C, Hassan AHS, Lehrberger K et al. Local analgesic effect of endogenous opioid peptides. Lancet 1993; 342:321-324.
25. Stein C, Lehrberger K, Yassouridis A et al. Opioids as novel intra-articular agents in arthritis. Prog Pain Res Manage 1994; 1:289-296.
26. Lee SH, Kayser V, Guilbaud G. Antinociceptive effect of systemic kelatorphan, in mononeuropathic rats, involves different opioid receptor types. Eur J Pharmacol 1994; 264:61-67.
27. Horwell DC. Kappa opioid analgesics. Drugs Future 1988; 13:1061-1070.
28. Barber A, Gottschlich R. Central & Peripheral Nervous Systems: Novel developments with selective, nonpeptidic kappa-opioid receptor agonists. Exp Opin Invest Drugs 1997; 6:1351-1368.
29. Barber A, Bartosyk GD, Bender HM et al. A pharmacological profile of the novel, peripherally-selective kappa-opioid receptor agonist, EMD 61753. Br J Pharmacol 1994; 113:1317-1327.
30. Rogers H, Bgirch PJ, Harrison SM et al. GR94839, kappa-opioid agonist with limited access to the central nervous system, has antinociceptive activity. Br J Pharmacol 1992; 106:783-789.
31. Shaw JS, Carroll JA, Alock P et al. ICI 204448: a kappa-opioid agonist with limited access to the CNS. Br J Pharmacol 1989; 96:986-992.
32. Birch PJ, Rogers H, Hayes AG et al. Neuroprotective actions of GR89696, a highly potent and selective κ-opioid receptor agonist. Br J Pharmacol 1991; 103:1819-1823.
33. Wilson JL, Nayanar V, Walker JS. The site of anti-arthritic action of the kappa agonist, U50488H: Importance of local administration. Br J Pharmacol 1996; 118:1754-1760.
34. Binder W, Carmody J, Walker J. Effect of gender on anti-inflammatory and analgesic actions of two kappa-opioids. J Pharmacol Exp Ther 2000; 292:303-309.
35. Binder W, Scott C, Walker JS. Involvement of substance P in the anti-inflammatory effects of the peripherally selective κ-opioid, asimadoline and the NK1 antagonist GR205171. Eur J Neurosci 1999; 11:2065-2072.
36. Wilson JL. PhD thesis submitted to the University of New South Wales 1998.
37. Pincus T. Rheumatoid arthritis: a medical emergency? Scand J Rheum1994; 23:21-30.
38. Giles BE, Walker JS. Gender differences in pain. Curr Op Anaesthesiol 1999; 12:591-595.
39. Cicero TJ, Nock B, Meyer ER. Gender-related differences in the antinociceptive properties of morphine. J Pharm Exp Ther 1996; 279:767-773.
40. Gear RW, Miaskowski NC, Gordon NC et al. Kappa-opioids produce significantly greater analgesia in women than in men. Nature Med 1996; 2:1248-1250.

41. Sharp BM, Roy S, Bidlack JM. Evidence for opioid receptors on cells involved in host defence and the immune system. J Neuroimmunol 1998; 83:45-46.
42. Morand EF, Cooley H, Leech M et al. Advances in the understanding of neuroendocrine function in rheumatic disease. Aust NZ J Med 1996; 26:543-551.
43. Hall J, Morand EF, Medbak S et al. Abnormal hypothalamic-pituitary-adrenal axis function in rheumatoid arthritis. Arthritis Rheum 1994; 8:1132-1137.
44. Bateman A, Singh A, Kral T et al. The immune-hypothalamic-pituitary-adrenal axis. Endocrine Reviews 1989; 10:92-112.
45. Blalock JE. A molecular basis for bidirectional communication between the immune and neuroendocrine systems. Physiol Rev 1989; 69:1-32.
46. Perretti M, Becherucci C, Scapigliati G et al. The effect of adrenalectomy on interleukin-1 release in vitro and in vivo. Br J Pharmacol 1989; 98:1137-1142.
47. Kidd BI, Mapp PI, Gibson SJ et al. A neurogenic mechanism for symmetrical arthritis. Lancet ii, 1989; 1128-1130.
48. Schaible HG, Grubb BD. Afferent and spinal mechanisms of joint pain. Pain 1993; 55:5-54.
49. Walker JS, Scott C, Bush KA et al. Effects of the peripherally selective κ-opioid asimadoline, on substance P and CGRP mRNA expression in chronic arthritis of the rat. Neuropeptides 2000; 34:193-202.
50. Carr DJJ. The role of endogenous opioids and their receptors in the immune system. Proc Soc Exp Biol Med 1991; 198:710-720.
51. Stein C. The control of pain in peripheral tissue by opioids. N Engl J Med 1995; 332:1685-1690.
52. Butcher EC, Picker LJ. Lymphocyte homing homeostasis. Science 1996; 272:60-66.
53. Machelska H, Cabot PJ, Mousa SA et al. Pain control in inflammation governed by selectins. Nature Med 1998; 4:1425-1428.
54. Szekanecz Z, Haines GK, Lin TR et al. Differential distribution of intercellular adhesion molecules (ICAM-1, ICAM-2 and ICAM-3) and the MS-1 antigen in normal and diseased human synovia. Their possible pathogenetic and clinical significance in rheumatoid arthritis. Arthritis Rheum 1994; 37:221-231.
55. Wilson JL, Walker JS, Antoon JS et al. Intercellular adhesion molecule-1 expression in adjuvant arthritis in rats: Inhibition by kappa-opioid agonist but not by NSAID. J Rheumatol 1998; 25:499-505.
56. Mikecz K, Brennan FR, Kim JH et al. The role of adhesion molecules in the development of autoimmune arthritis, Scand J Rheumatol 1995; 24:99-106.
57. Ohh M, and Takei F. New insights into the regulation of ICAM-1 gene expression. Leuk. Lymphoma 1995; 20:223-8.
58. Belkowski SM, Alicea C, Eisenstein TK et al. Inhibition of IL-1 and TNF-alpha synthesis following treatment of macrophages with the kappa opioid agonist U-50,488H. J Pharmacol Exp Ther 1995; 273:1491-1496.
59. Issekutz AC, Ayer L, Miyasaka M et al. Treatment of established adjuvant arthritis in rats with monoclonal antibody to CD18 and very late activation antigen-4 integrins suppresses neutrophil and T-lymphocyte migration to the joints and improves clinical disease. Immunology 1996; 88:569-576.
60. Wahl SM, Hunt DA, Bansal G et al. Bacterial cell wall-induced immunosuppression: role of transforming growth factor β. J. Exp Med 1988; 168:1403-1417.

Index

Symbols

β-endorphin 40-42, 44, 69, 77, 78, 80-86, 89, 90, 92-95, 98, 99, 106, 109-113, 119, 127, 130, 131
δ opioid receptor 98
κ-agonist 72, 101, 133, 134, 148, 149, 153-155
μ-receptors 71, 79, 89, 110, 132, 140

A

ACTH 41, 46, 53, 55-62, 119, 127, 130, 131, 135, 136, 141, 153
Adhesion molecules 60, 77, 82, 83, 88, 90, 91, 96, 154, 155
Allodynia 1, 3, 6, 7, 10-12, 14, 15
Analgesia 1, 2, 12, 22, 32, 40, 43-48, 51, 53, 69, 70, 72, 73, 77, 78, 81, 82, 88-90, 93-96, 98, 106, 107, 117, 118, 122, 126, 140, 148, 154
Analgesic cytokines 29, 34
Anti-inflammatory actions 24, 32, 71, 149, 153, 154, 156, 157
Antisense treatment 46
Astrocytes 2-9, 11-13

B

Brain 1-5, 10-15, 21, 25, 32, 43, 52, 53, 59, 61, 70, 79, 89, 106, 109, 111, 127, 132, 149

C

CD10 136
Clinical implications 5, 62, 95
Corticotropin 41, 42, 44-46, 52-54, 62, 77, 81, 82, 88, 93, 94, 109, 119, 127, 130, 131, 135
Corticotropin-releasing factor (CRF) 41-48, 55, 57, 60-62, 77, 81-83, 88, 91-94, 96, 119, 153
Cytokines 1, 2, 4-8, 10-15, 22-34, 40, 41, 47, 48, 59-61, 77, 83, 88, 91, 93, 98, 107, 111, 113, 118, 129, 136, 137, 141, 153, 155-158
Cytokine release 13, 15, 23, 27, 28, 30, 31, 155

E

Endogenous analgesia 93
Endorphin 40-42, 44, 51-54, 57, 59, 62, 63, 69, 77, 78, 80-85, 89, 90, 92-95, 98, 99, 106, 109-113, 119, 127, 129-131
Endotoxic shock 59, 60
Enkelytin 128, 129, 131
Enkephalin 40, 44, 45, 51-53, 58, 60, 62, 63, 69, 78-81, 83, 89, 93-95, 98, 99, 102, 106, 109, 127-136
Experimental arthritis 153
Extraterritorial allodynia 11

F

Fluorocitrate 6, 11, 14

G

Glia 1-6, 13, 14
Glial fibrillary acidic protein (GFAP) 3, 7-9
gp120 (envelope protein of the HIV-1 virus) 5-8, 14, 61

H

Hyperalgesia 1-3, 5-7, 10-12, 14, 22-34, 40, 43, 88, 118, 122, 150
Hyperalgesia memory 31
Hyperalgesic cytokines 22, 23, 25-27, 29, 31, 32, 34

I

Immune 1-5, 7-12, 14, 22, 40-43, 45-48, 51-53, 56-58, 60-62, 69, 71, 72, 77-84, 88-93, 95, 96, 98-100, 102, 103, 106-113, 117, 118, 126, 127, 129-131, 133, 135-138, 140, 141, 148, 151-155, 157
Immune cells 2, 3, 5, 14, 40-43, 45-48, 51, 62, 69, 71, 77-83, 88-91, 93, 95, 96, 98-100, 103, 106, 109, 111-113, 118, 133, 138, 141, 143, 151, 153-155
Immunohistochemistry 3, 7, 8, 77, 80, 150
Immunomodulation 52, 106, 107, 109, 141
Inflammation 3-5, 11, 13, 22, 23, 28, 29, 31, 40, 42, 43, 48, 58, 60-62, 71, 72, 77-83, 88-94, 96, 110, 111, 118, 149, 151-155
Inflammatory hyperalgesia 22, 23, 25, 28, 30, 31, 33, 34
Inflammatory pain 23, 26-28, 34, 40, 43, 70, 77- 79, 83, 89, 90
Integrins 82, 90, 91
Interleukin 1, 4, 6, 8, 9, 23, 26, 27, 31, 40, 42, 44-46, 55, 62, 81, 82, 92, 93, 98, 107, 109, 111, 113, 118, 119, 132, 137, 155, 158
Interleukin (IL)-1 4-9, 11-14, 24, 27, 29-31, 33, 41-48, 55, 57-62, 81, 82, 92, 93, 119, 132, 137, 155, 156
Interleukin (IL)-6 1, 4, 5, 12-14, 24-27, 29, 33, 34, 40, 41, 59, 111, 118, 122
Interleukin-1 receptor antagonist (IL-1ra) 6, 28-31, 34, 93
Intrinsic pain control 47
Invertebrate 126-141

L

Lymphocytes 23, 29-31, 41-43, 51-60, 61-63, 79-83, 89-92, 95, 99-102, 106-111, 117, 118, 136, 140, 154
Lymphoproliferation 109, 110

M

MAPK 102
Matrix metalloproteinases 13, 14
Mechanical allodynia 1, 3, 6, 7, 10-12, 14, 15
Microglia 2-7, 11, 13, 129
Mirror pain 10
Mitogen activated protein 6
Morphine 32, 34, 51-53, 70-73, 95, 96, 98, 99, 102, 106-110, 113, 120-122, 126, 132-141, 149-151, 155
Morphine-6-glucoronide 138

N

Naloxone 46, 52, 53, 59, 82, 93, 95, 99, 107, 110-113, 127, 132-134, 136, 137, 140, 141, 149, 151
Natural killer (NK) cells 53, 63, 107-110, 117-121
Neuroimmune 137
NK activity 108-110, 117, 119

O

Opiate 2, 51-53, 59, 63, 98, 99, 106-108, 112, 113, 126, 127, 131, 132, 134, 136-141
Opioid peptide 40, 41, 43-48, 51, 52, 60, 62, 69, 71, 72, 74, 77-84, 88-91, 93-98, 106, 109, 111-113, 127-129, 131, 133, 134, 136, 137, 140, 141, 148, 154
Opioid receptors 40, 41, 43, 44, 45, 46, 51-53, 58, 59, 62, 63, 69, 70, 71, 72, 73, 77, 78, 79, 82, 83, 88, 89, 93, 95, 96, 98, 99, 100-103, 106, 107, 109, 110, 111, 114, 115, 127, 131-134, 136, 137, 140, 148, 149, 150, 151, 152, 154, 157
Opioids 40, 41, 43-48, 51-53, 55, 56, 58-63, 69-73, 77-84, 88-91, 93, 94, 95, 96, 98-100, 102, 103, 106, 107, 109-113, 121, 126-129, 131, 132, 133-137, 140, 141, 148-152, 154, 155-157
OX42 7

P

p38 MAP kinase 7, 11, 14
p38 MAP kinase inhibitor 11, 14
Pentoxifylline 14, 15, 33
Peripheral analgesia 40, 78, 154
Peripheral CRF receptors 92
Peripheral hyperalgesia 24
Peripheral opioid analgesia 82, 90, 93, 94, 154
Peripheral receptors 89
Proinflammatory cytokines 1, 2, 4-8, 11-15, 59
Proliferation 58, 60, 61, 63, 98, 99, 103, 107-111, 140, 154
Proopiomelanocortin (POMC) 41, 51-53, 55-62, 78-80, 119, 130, 131, 135, 141

R

Receptors 5, 6, 10-13, 22, 23, 28, 30, 40-46, 48, 51-53, 57-63, 69, 70-73, 77-79, 81-84, 88, 89, 92, 93, 95, 96, 98-100, 103, 106, 109, 110, 127, 129, 131-134, 136, 137, 141, 148-152, 154, 157
Rheumatoid arthritis 28, 83, 148, 154, 157

S

Sciatic inflammatory neuritis (SIN) 10, 11, 14
Sciatic nerve 9-11, 33, 70, 71, 77-79
Selectins 88, 89, 91, 93
Signal transduction 14, 109, 134
Skin graft 111
Spinal cord 1-7, 9-13, 15, 69, 70, 79
Stress-induced immunosuppression 111, 112
Surgery 10, 12, 72, 83, 95, 117-123, 131, 135, 149

T

T-cells 55-58, 79, 81, 90, 98-103, 107, 108, 140, 156
Thl/Th2 111, 113
Thalidomide 13, 26, 27, 32, 33
Thermal hyperalgesia 1-3, 5-7, 10-12, 14, 24, 25, 27, 28, 31, 118
Transcription 55, 56, 100, 101, 131, 156
Tumor 1, 4, 24, 26, 27, 40, 53, 57, 113, 117-123, 137, 155, 157, 158
Tumor necrosis factor (TNF) 1, 4, 5, 7, 8, 11-14, 24-34, 40, 41, 59, 60, 113, 137, 140, 155-158

Z

Zymosan 3, 4, 10, 11, 24, 28, 33

www.ingramcontent.com/pod-product-compliance
Lightning Source LLC
Chambersburg PA
CBHW080735150926
51501CB00005B/173
* 9 7 8 0 3 0 6 4 7 6 9 2 1 *